# Edexcel International GCSE Chemistry

# Edexcel Certificate in Chemistry

## Revision Guide

Cliff Curtis

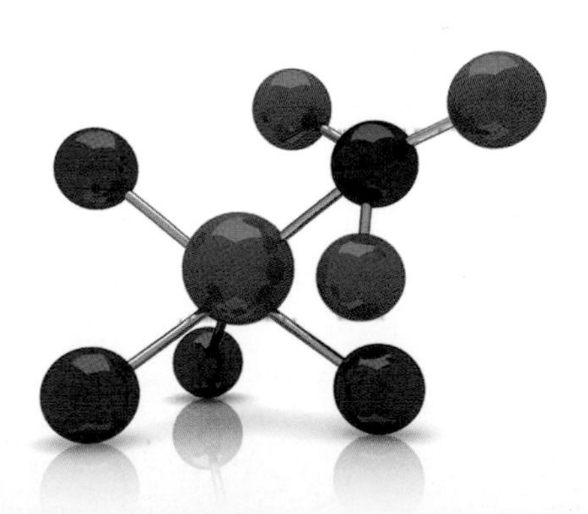

ALWAYS LEARNING

PEARSON

Published by Pearson Education Limited, a company incorporated in England and Wales, having its registered office at Edinburgh Gate, Harlow, Essex, CM20 2JE. Registered company number: 872828.

www.pearsonglobalschools.com

Edexcel is a registered trademark of Edexcel Limited



First published 2011

20 19 18 17 16 15

IMP 20 19 18 17 16 15 14 13 12 11 10

ISBN 978 0 435046 72 9



Original design by Richard Ponsford

Edited by Alexandra Clayton

Proofread by Ralph Footring

Packaged and typeset by Naranco Design & Editorial / Kim Hubbeling

Original illustrations © Pearson Education Ltd 2011

Illustrated by Andriy Yankovsky

Cover design and title page by Creative Monkey

Cover photo © iStockphoto: Pavlen

Printed in China (CTPS/10)

**Acknowledgements**
The author and publisher would like to thank the following individuals and organisations for permission to reproduce photographs:

(Key: b-bottom; c-centre; l-left; r-right; t-top)

Alamy Images: 1l; Fotolia.com: 1r; iStockphoto: Pavlen header; Pearson Education Ltd: 3cl, 3cr, 3bl, 3br, 43r, 50br, 69br, 80bl, 81tr, 81cr; Science Photo Library Ltd: 3tl, 3tr, 49br, 69t, 69c, 69b, 101tl, 101bl; Shutterstock.com: 1c

Cover images: iStockphoto: Pavlen br

All other images © Pearson Education 2011

Every effort has been made to contact copyright holders of material reproduced in this book. Any omissions will be rectified in subsequent printings if notice is given to the publishers.

**Websites**
The websites used in this book were correct and up to date at the time of publication. It is essential for tutors to preview each website before using it in class so as to ensure that the URL is still accurate, relevant and appropriate. We suggest that tutors bookmark useful websites and consider enabling students to access them through the school/college intranet.

Contents

# About this revision guide

The aim of this revision guide is to help you succeed in your Edexcel International GCSE Chemistry examinations.

The guide covers the Edexcel International GCSE Chemistry Specification 4CHO. It is divided into seven main sections.

The first five sections cover the topics in order of the specification.

Section F offers advice on how to answer questions assessing practical skills. The terms that will be used in these types of questions are explained so that you will be aware of the answers that the examiners will be expecting.

Section G offers hints on answering examination questions, particularly those that require you to apply your knowledge and understanding of Chemistry. You will see examples of the sorts of mistakes that candidates often make and how to avoid these.

## Examination questions

At the end of each of the first six sections, you will find examination questions taken from past Edexcel International GCSE Chemistry papers. Although these questions are taken from the previous specification, they have been carefully chosen to make sure that they are relevant to the current specification.

The answers to these questions, together with further explanations as necessary, appear on the CD.

## Using the revision guide

Revision is a very important part of your preparation for the examination.

- Make sure you leave yourself with plenty of time to revise.
- Revise in several short sessions rather than one long one. Your brain needs to rest between sessions, so do something relaxing like taking a walk or listening to classical music.
- Plan which section of the book to cover in each revision session.
- Include time in your plan for answering past examination questions. This is an excellent way to reinforce what you have learnt.
- Make a note of anything that you are unsure of and talk to your teacher about this as soon as you can.

I hope this book helps you to do well in your International GCSE Chemistry examinations.

Cliff Curtis

# Chapter 1: States of matter

- There are three states of matter: solid, liquid and gas.
- All matter is made up of tiny particles that are too small to be seen with the naked eye.
- The particles are arranged differently in each state of matter. They also have different types of movement.
- Solids, liquids and gases can be changed into one another by changing the amount of energy that the particles have.

| | Solid | Liquid | Gas |
|---|---|---|---|
| State of substance | example: ice | example: water | example: water vapour |
| Particles in substance | solid | liquid | gas |
| Spacing between particles | close together and touching one another | close together and touching one another | far apart |
| Arrangement of particles | regular, repeating pattern | irregular | irregular |
| Movement of particles | vibrate about fixed positions but do not move apart | move around and slide past one another | move freely and constantly collide with each other |
| Forces between particles | stronger than in the liquid | not as strong as in the solid | non-existent |

## Changes of state

| Change | How to bring about the change | Why the change occurs |
|---|---|---|
| Solid to liquid | Heat the solid until it melts | The particles **gain kinetic energy** and vibrate faster and faster. This allows the particles to **overcome** the forces of attraction that hold them together in the solid. The regular pattern is broken down and the particles can now **slide past** one another |
| Liquid to solid | Cool the liquid until it freezes | The particles **lose kinetic energy** and this allows the forces of attraction between the particles to **hold them together**. The particles arrange themselves into a **regular pattern** and are no longer able to slide past one another |
| Liquid to gas | Heat the liquid until it boils | The particles **gain kinetic energy** and move further apart. Eventually the forces of attraction between the particles are **completely broken** and they are able to **escape** from the liquid |
| Gas to liquid | Cool the gas until it condenses | The particles **lose kinetic energy** and this allows the forces of attraction to bring the particles **closer together**. The particles eventually **clump together** to form a liquid |
| Solid to gas | Heat the solid until it sublimes | The particles **gain kinetic energy** and vibrate faster and faster. Eventually the forces of attraction between the particles are **completely broken** and they are able to **escape** from the solid |

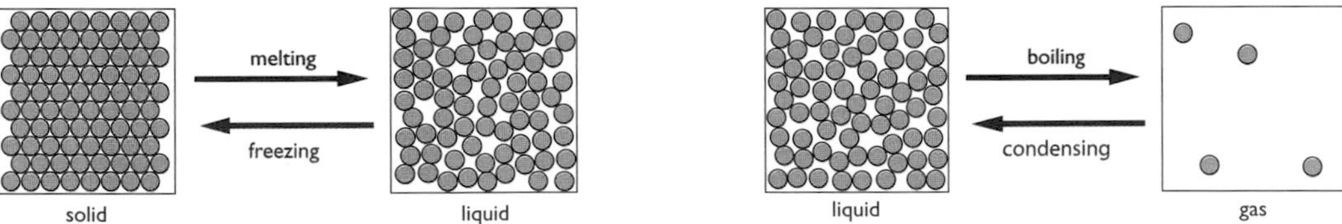


**Figure 1.1** *Melting to become a liquid and freezing to become a solid; boiling to become a gas and condensing to become a liquid.*

# Chapter 2: Atoms

How do we know that matter is made up of tiny, moving particles? The true answer to this question is that we do not know for certain. However, there is a lot of evidence to support this theory.

## Dilution of coloured solutions

When potassium manganate(VII) crystals are dissolved in water, a purple solution is formed. A few very tiny crystals can produce a highly intense colour.

When this solution is diluted several times, the colour fades, but does not disappear until a lot of dilutions are made.

This indicates that there are a large number of particles of potassium manganate(VII) in a very small amount of solid. If this is true, then the particles of potassium manganate(VII) must be very tiny.

## Diffusion

Particles will move to fill the space available to them. They can do this in both liquids and gases.

An example is the **diffusion** of bromine from one flask to another. After five minutes the bromine gas has diffused into the left-hand flask. This happens because both air and bromine particles are moving randomly and there are large gaps between the particles. The particles can therefore easily mix together.

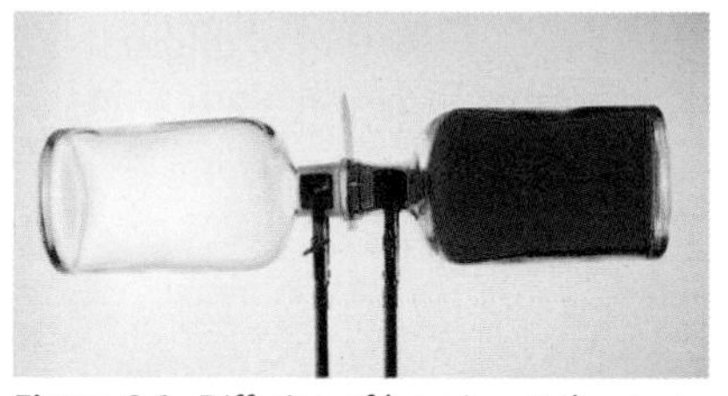

**Figure 2.1** *Diffusion of bromine at the start...*

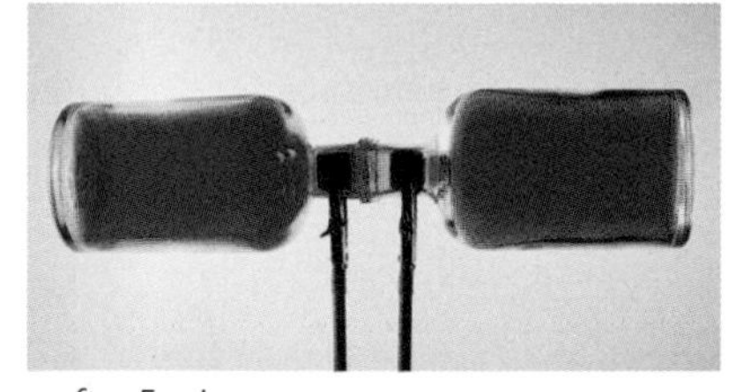

*...after 5 mins.*

Diffusion also takes place in liquids, only more slowly since the particles in a liquid are closely packed together and they move more slowly than in a gas.

The photos below show potassium manganate(VII) (potassium permanganate) first of all dissolving and then diffusing in water.

**Figure 2.2** *Diffusion of potassium manganate(VII) at the start*

*after 2 hours*

*after 4 hours*

*after 6 hours*

## Atoms and molecules

**Atoms** are made up of smaller, sub-atomic particles called protons, neutrons and electrons ***(see Chapter 3)***.

Most **molecules** are made up of two or more atoms covalently bonded together *(see Chapter 7)*. However, the noble gases (helium, neon, argon, krypton, xenon and radon) exist as atoms at room temperature and pressure, so for them the atom and the molecule are the same. Molecules that contain only one atom are called **monatomic molecules**.

## Elements, compounds and mixtures

**Elements** and **compounds** have their own chemical properties. The properties of a compound are likely to be different from the elements that have been used to make it. A **mixture** will have the properties of each substance that has been used to make it.

This information is summarised in the table below:

| Name of substance | Type of substance | Appearance | Action of a magnet | Action of dilute hydrochloric acid |
|---|---|---|---|---|
| Iron (filings) | Element | Grey powder | Attracted | Reacts to form hydrogen gas |
| Sulfur | Element | Yellow powder | Not attracted | Does not react |
| Iron + sulfur | Mixture | Mixture of grey and yellow powders | Grey powder (iron) attracted<br><br>Yellow powder (sulfur) not attracted | Grey powder reacts to form hydrogen gas ($H_2$)<br><br>Yellow powder does not react |
| Iron(II) sulfide | Compound | Dark grey solid | Not attracted | Reacts to form hydrogen sulfide gas ($H_2S$) |

### Separation of mixtures

Since the components of a mixture are not chemically bonded together, they can be separated without carrying out any chemical reactions.

Five techniques for separating mixtures are described in the table below:

| Name of separation technique | What the technique is used to do |
|---|---|
| Filtration | To separate an undissolved solid from a mixture of the solid and a liquid/solution (e.g. sand from a mixture of sand and water) |
| Evaporation | To separate a dissolved solid from a solution, when the solid has similar solubilities in both cold and hot solvent (e.g. sodium chloride from a solution of sodium chloride in water) |
| Crystallisation | To separate a dissolved solid from a solution, when the solid is much more soluble in hot solvent than in cold (e.g. copper(II) sulfate from a solution of copper(II) sulfate in water) |
| Simple distillation | To separate a liquid from a solution (e.g. water from a solution of sodium chloride in water) |
| Fractional distillation | To separate two or more liquids that are miscible with one another (e.g. ethanol and water from a mixture of the two) |
| Paper chromatography | To separate substances that have different solubilities in a given solvent (e.g. different coloured inks that have been mixed to make black ink) |

## Separation of compounds

Since the elements in a compound are chemically joined together, they can only be separated from one another by carrying out a chemical reaction.

Sometimes heating the compound will be sufficient to make that reaction happen. For example, mercury oxide decomposes (breaks down) into mercury and oxygen on heating:

$$2HgO \rightarrow 2Hg + O_2$$

This is known as **thermal decomposition**.

On other occasions, electricity will decompose the compound. For example, if an electric current is passed through molten lead(II) bromide, it breaks down into lead and bromine.

This is known as **electrolysis** *(see Chapter 9)*.

# Chapter 3: Atomic structure

Atoms are made up of sub-atomic particles called **protons, neutrons** and **electrons**.

The only atom that does not contain any neutrons is the simplest **isotope** of hydrogen *(see the table on isotopes on page 6)*.

The **nucleus** of the atom contains protons and neutrons. It is shown highly magnified in this diagram.

In reality, if you scaled a helium atom up to the size of a sports hall, the nucleus would be no more than the size of a grain of sand.

The **electrons** are found at large distances from the nucleus. In this case, they are found most of the time somewhere in the shaded red area.

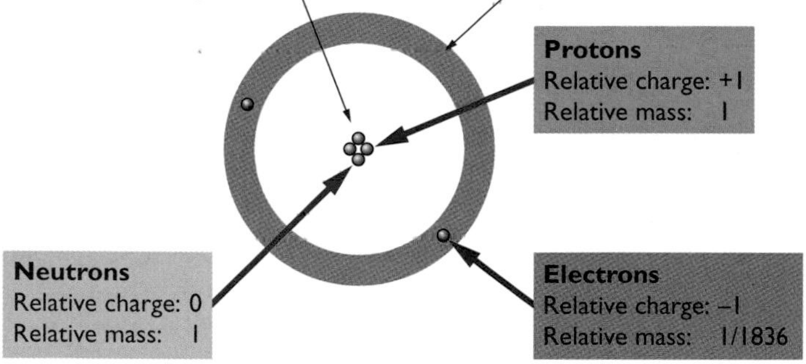

**Protons**
Relative charge: +1
Relative mass: 1

**Neutrons**
Relative charge: 0
Relative mass: 1

**Electrons**
Relative charge: –1
Relative mass: 1/1836

**Figure 3.1** *The structure of a helium atom.*

The electrons have virtually no mass compared with the masses of protons and neutrons; therefore nearly all of the mass of the atom is concentrated in the nucleus.

## Atomic number, mass number and isotopes

Atomic number = number of protons in the nucleus of an atom of an element

Mass number = number of protons + number of neutrons in the nucleus of an atom of an element

For any atom, this information can be shown simply as:

$$^{35}_{17}Cl$$

mass number → 35; atomic number → 17; Cl ← symbol for element

This atom of chlorine contains 17 protons and 18 neutrons (17 + 18 = 35 = mass number).

Isotopes are atoms that have the same atomic number but different mass numbers. This is because they contain the **same** number of **protons** but **different** numbers of **neutrons**.

The composition of the three isotopes of hydrogen is given in the table:

| Isotope | Number of protons | Number of neutrons | Symbol for isotope |
|---|---|---|---|
| Hydrogen-1 | 1 | 0 | $^{1}_{1}H$ |
| Hydrogen-2 | 1 | 1 | $^{2}_{1}H$ |
| Hydrogen-3 | 1 | 2 | $^{3}_{1}H$ |

# The electrons

## Counting the number of electrons in an atom

The number of electrons in an atom is the same as the number of protons. Hence, all of the hydrogen atoms listed in the table above contain 1 electron.

The atomic number of an element can be found in the Periodic Table *(see page 119)*.

All atoms of carbon (atomic number = 6) contain 6 electrons, whilst all atoms of sodium (atomic number = 11) contain 11 electrons.

## How the electrons are arranged in atoms

Electrons exist **around** the nucleus of atoms in different **energy levels** called **electron shells**. There are several **electron shells**.

The first electron shell surrounds the nucleus. The second surrounds the first and is therefore **bigger** and **further away** from the nucleus. The third electron shell surrounds the second and is therefore **bigger** than both the first and the second electron shells and **further away** from the nucleus than both of these. You can view the shells as being like the layers around an onion.

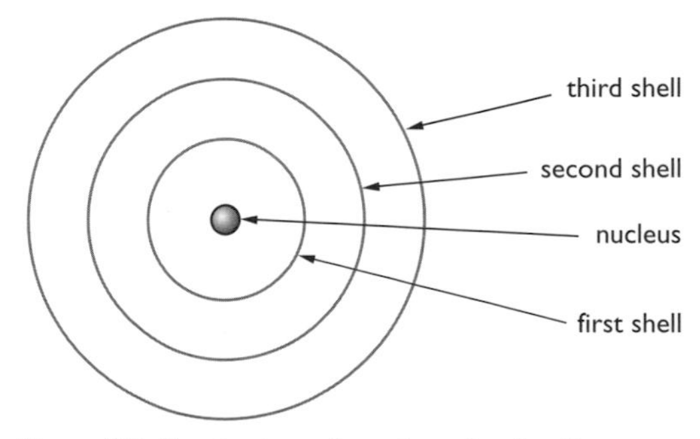


**Figure 3.2** *The structure of an atom showing the arrangement of electron shells.*

Each electron shell can accommodate a fixed number of electrons:

- the **first** shell can take up to a maximum of **2** electrons
- the **second** shell can take up to a maximum of **8** electrons
- the **third** shell can take up to a maximum of **18** electrons.

When you are trying to work out in which shells the electrons of a particular atom will be, the following rules should be followed:

**1.** Electrons always occupy those shells that are **closest** to the nucleus.

**2.** When a shell becomes full, the next one is started.

The arrangement of electrons into shells for an atom is known as its **electronic configuration**.

Using these two rules, the electronic configurations for the first 20 elements are shown in the table below:

| Element | Number of electrons | 1st shell | 2nd shell | 3rd shell | 4th shell |
|---|---|---|---|---|---|
| Hydrogen | 1 | 1 | | | |
| Helium | 2 | 2 | | | |
| Lithium | 3 | 2 | 1 | | |
| Beryllium | 4 | 2 | 2 | | |
| Boron | 5 | 2 | 3 | | |
| Carbon | 6 | 2 | 4 | | |
| Nitrogen | 7 | 2 | 5 | | |
| Oxygen | 8 | 2 | 6 | | |
| Fluorine | 9 | 2 | 7 | | |
| Neon | 10 | 2 | 8 | | |
| Sodium | 11 | 2 | 8 | 1 | |
| Magnesium | 12 | 2 | 8 | 2 | |
| Aluminium | 13 | 2 | 8 | 3 | |
| Silicon | 14 | 2 | 8 | 4 | |
| Phosphorus | 15 | 2 | 8 | 5 | |
| Sulfur | 16 | 2 | 8 | 6 | |
| Chlorine | 17 | 2 | 8 | 7 | |
| Argon | 18 | 2 | 8 | 8 | |
| Potassium | 19 | 2 | 8 | 8 | 1 |
| Calcium | 20 | 2 | 8 | 8 | 2 |

The electronic configurations of potassium and calcium are not as expected. In both cases the outermost electrons are in the 4th shell instead of the 3rd shell, despite the fact that the 3rd shell can accommodate up to 18 electrons. The reasons for this are very complex and are covered at AS level.

The number of electrons in the outermost shell is the same as the Group number for the element in the Periodic Table. For example, lithium, sodium and potassium all have 1 electron in their outermost shell and are in Group 1 of the Periodic Table.

Electronic configurations are often given in a shorthand notation. For example:

Carbon, C 2.4 Magnesium, Mg 2.8.2 Potassium, K 2.8.8.1

## Relative atomic mass

The **relative atomic mass** is calculated from the masses and relative abundances of all the isotopes of a particular element. It is usually given the symbol $A_r$.

**Worked Example**

Chlorine has two isotopes: chlorine-35 and chlorine-37

A typical sample of chlorine will be 75% chlorine-35 atoms and 25% chlorine-37 atoms.

The relative atomic mass is calculated as follows:

Total mass of 100 atoms $= (75 \times 35) + (25 \times 37) = 3550$

Mean mass of one atom $= (3550 \div 100) = 35.5$

**$A_r$ of chlorine is 35.5**

Most elements have isotopes and hence their relative atomic masses will not be a whole number. However, for the sake of simplicity, most relative atomic masses quoted in the Periodic Table are given to the nearest whole number.

**Worked Example 2**

The isotopes of magnesium and their percentage abundances are:

Magnesium-24 78.6%; magnesium-25 10.1%; magnesium-26 11.3%

Total mass of 100 atoms $= (78.6 \times 24) + (10.1 \times 25) + (11.3 \times 26) = 2432.7$

Mean mass of one atom $= (2432.7 \div 100) = 24.327$

**$A_r$ of magnesium is 24.3 (to one decimal place)**

# Chapter 4: Relative formula mass

## Calculating relative formula masses

Relative formula mass is given the symbol $M_r$.

To calculate the $M_r$ of a substance, all you have to do is add up the relative atomic masses of all the atoms present in the formula.

**Worked Examples**

| Substance | Atoms present | $M_r$ |
|---|---|---|
| Hydrogen, $H_2$ | $2 \times H$ | $(2 \times 1) = 2$ |
| Water, $H_2O$ | $(2 \times H) + (1 \times O)$ | $(2 \times 1) + 16 = 18$ |
| Potassium carbonate, $K_2CO_3$ | $(2 \times K) + (1 \times C) + (3 \times O)$ | $(2 \times 39) + 12 + (3 \times 16) = 138$ |
| Calcium hydroxide, $Ca(OH)_2$ | $(1 \times Ca) + (2 \times O) + (2 \times H)$ | $40 + (2 \times 16) + (2 \times 1) = 74$ |
| Ammonium sulfate, $(NH_4)_2SO_4$ | $(2 \times N) + (8 \times H) + (1 \times S) + (4 \times O)$ | $(2 \times 14) + (8 \times 1) + 32 + (4 \times 16) = 132$ |

Most candidates do not have a problem calculating relative formula masses until they meet an example involving water of crystallisation.

**Worked Example**

Hydrated magnesium sulfate, $MgSO_4.7H_2O$

In cases like this, calculate the $M_r$ of $MgSO_4$ and then add 7 × the $M_r$ of $H_2O$. Hence,

$MgSO_4 = 24 + 32 + (4 \times 16) = 120$

$7H_2O = (7 \times 18) = 126$

**$M_r$ of $MgSO_4.7H_2O$ = (120 + 126) = 246**

## The mole

The **mole** is a measure of the **amount of substance**.

One mole (1 mol) is the amount of substance that contains $6 \times 10^{23}$ particles (atoms, molecules, or formulae) of the substance.
$6 \times 10^{23}$ is known as the **Avogadro number**.

For example:

1 mol of sodium (Na) contains $6 \times 10^{23}$ **atoms** of sodium

1 mol of hydrogen ($H_2$) contains $6 \times 10^{23}$ **molecules** of hydrogen

1 mol of sodium chloride (Na*Cl*) contains $6 \times 10^{23}$ **formulae** of sodium chloride

Sodium chloride is an ionic compound and therefore does not contain any molecules. This is why it is important to refer to a certain number of **formulae**, and not **molecules**, being present *(see Chapters 6 and 7)*.

### Calculating the mass of one mole

The mass of one mole of atoms is easily calculated. It is simply the relative atomic mass ($A_r$) expressed in grams.

**Worked Examples**

| Element | Symbol | $A_r$ | Mass of one mole of atoms |
|---|---|---|---|
| Hydrogen | H | 1 | 1 g |
| Carbon | C | 12 | 12 g |
| Oxygen | O | 16 | 16 g |
| Sodium | Na | 23 | 23 g |
| Chlorine | *Cl* | 35.5 | 35.5 g |

It is important to say precisely what you are talking about when quoting the mass of a mole of something, otherwise it might lead to confusion.

For example, 1 mol of hydrogen could mean:

- 1 mol of hydrogen atoms, H, mass 1 g

*or*

- 1 mol of hydrogen molecules, $H_2$, mass 2 g

Similarly, 1 mol of magnesium sulfate could mean:

- 1 mol of anhydrous magnesium sulfate, $MgSO_4$, mass 120 g

*or*

- 1 mol of hydrated magnesium sulfate, $MgSO_4.7H_2O$, mass 246 g

The mass of one mole of a substance is also easily calculated by working out the formula mass ($M_r$) and expressing it as grams.

**Worked Examples**

| Substance | Formula | $M_r$ | Mass of one mole of molecules/ formulae |
|---|---|---|---|
| Hydrogen | $H_2$ | 2 | 2 g |
| Oxygen | $O_2$ | 32 | 32 g |
| Water | $H_2O$ | 18 | 18 g |
| Sodium chloride | NaCl | 58.5 | 58.5 g |
| Hydrated magnesium sulfate | $MgSO_4.7H_2O$ | 246 | 246 g |

## Simple calculations with moles

### *1. Calculating mass from amount (i.e. number of moles)*

Mass of substance (in grams) = amount × $M_r$

**Worked Examples**

| Substance | Amount | $M_r$ | Mass |
|---|---|---|---|
| $H_2O$ | 0.5 mol | 18 | (0.5 × 18) = **9 g** |
| NaCl | 3 mol | 58.5 | (3 × 58.5) = **175.5 g** |
| $K_2CO_3$ | 0.2 mol | 138 | (0.2 × 138) = **27.6 g** |
| $(NH_4)_2SO_4$ | 2.5 mol | 132 | (2.5 × 132) = **330 g** |
| $MgSO_4.7H_2O$ | 0.25 mol | 246 | (0.25 × 246) = **61.5 g** |

### *2. Calculating amount from mass*

Amount in moles = mass of substance in grams ÷ $M_r$

**EXAMINER'S TIP**

Make sure the mass is in grams. If it is kilograms or tonnes, make sure you convert to grams before you perform your calculation.

**Worked Examples**

| Substance | Mass | $M_r$ | Amount |
|---|---|---|---|
| NaOH | 80 g | 40 | (80 ÷ 40) = **2 mol** |
| $CaCO_3$ | 25 g | 100 | (25 ÷ 100) = **0.25 mol** |
| $H_2SO_4$ | 4.9 g | 98 | (4.9 ÷ 98) = **0.05 mol** |
| $H_2O$ | 108 g | 18 | (108 ÷ 18) = **6 mol** |
| $CuSO_4.5H_2O$ | 75 g | 250 | (75 ÷ 250) = **0.3 mol** |
| $HNO_3$ | 126 kg | 63 | (126 000 ÷ 63) = **2000 mol** |

In the last example in the table, for $HNO_3$, 126 kg was converted to 126 000 g before the amount was calculated.

### 3. Calculations with atoms of elements

It is important to realise that the above calculations can be used with atoms of elements.

**Worked Examples**

0.5 mol of hydrogen atoms have a mass of $(0.5 \times 1) = 0.5$ g

32 g of oxygen atoms is equal to $(32 \div 16) = 2$ mol of oxygen atoms

These types of calculation are important when working out the empirical formulae of compounds ***(see Chapter 5).***

## Molar volume of a gas

One mole of any gas has a volume of 24 $dm^3$ (24 000 $cm^3$) at room temperature and pressure (r.t.p.). This is called the **molar volume** of a gas.

### 1. Calculating volume of gas from amount of gas

Volume of gas (in $\mathbf{dm^3}$) $=$ amount $\times$ 24,

**OR**

Volume of gas (in $\mathbf{cm^3}$) $=$ amount $\times$ 24 000

**Worked Examples**

| Name of gas | Amount of gas | Volume of gas |
|---|---|---|
| Hydrogen | 3 mol | $(3 \times 24) = 72\,dm^3$ |
| Carbon dioxide | 0.25 mol | $(0.25 \times 24) = 6\,dm^3$ |
| Oxygen | 5.4 mol | $(5.4 \times 24) = 129.6\,dm^3$ |
| Ammonia | 0.02 mol | $(0.02 \times 24) = 0.48\,dm^3$ |

### 2. Calculating amount of gas from the volume of gas

Amount of gas (in moles) $=$ volume of gas (in $\mathbf{dm^3}$) $\div$ 24,

**OR**

$=$ volume of gas (in $\mathbf{cm^3}$) $\div$ 24 000

**Worked Examples**

| Name of gas | Volume of gas | Amount of gas |
|---|---|---|
| Methane | $225.6\,dm^3$ | $(225.6 \div 24) = 9.4$ mol |
| Carbon monoxide | $7.2\,dm^3$ | $(7.2 \div 24) = 0.3$ mol |
| Sulfur dioxide | $960\,dm^3$ | $(960 \div 24) = 40$ mol |
| Oxygen | $1200\,cm^3$ | $(1200 \div 24000) = 0.05$ mol |

In the last calculation in the table, for $O_2$, the volume was given in $cm^3$ and therefore it was divided by 24 000, rather than by 24.

The calculation could also have been performed by converting the volume to $dm^3$ and then dividing by 24.

# Chapter 5: Chemical formulae and equations

## Writing balanced chemical equations

Balancing equations is not as hard as it first seems. It is simply a matter of getting the numbers correct. You must always have the same number of atoms of each element on either side of the equation.

- Work across the equation from left to right, checking one element after another, except if an element appears in several places in the equation. In that case, leave it until the end.
- If there is a group of atoms (such as a nitrate group, $NO_3$), which has not changed from one side to the other, then count the whole groups, rather than counting the individual atoms.
- Check everything at the end.

**Worked Example 1**

*Aluminium + copper(II) oxide → aluminium oxide + copper*

The unbalanced equation is $Al + CuO \rightarrow Al_2O_3 + Cu$

Count the aluminium atoms: 1 on the left; 2 on the right. If you end up with 2, you must have started with 2. The only way to achieve this is to have 2$Al$ (you must not write $Al_2$ since this is not the symbol for metallic aluminium).

The equation now reads $2Al + CuO \rightarrow Al_2O_3 + Cu$

Count the copper atoms: 1 on the left; 1 on the right. This is therefore okay for the moment, but be prepared to change this if you make other changes later.

Count the oxygen atoms: 1 on the left; 3 on the right. The only way of achieving 3 on the left is to have 3CuO (you cannot write $CuO_3$ since this is not the correct formula for copper(II) oxide).

The equation now reads $2Al + 3CuO \rightarrow Al_2O_3 + Cu$

Now count the copper atoms again: 3 on the left; 1 on the right. The only way of achieving 3 on the right is to have 3Cu (once again, you cannot have $Cu_3$, since this is not the symbol for copper).

The equation now reads $2Al + 3CuO \rightarrow Al_2O_3 + 3Cu$

Finally, count everything again to make sure.

$Al$ : 2 on the left; 2 on the right ✓

Cu: 3 on the left; 3 on the right ✓

O: 3 on the left; 3 on the right ✓

The equation is balanced.

**Worked Example 2**

*Magnesium oxide + nitric acid → magnesium nitrate + water*

The unbalanced chemical equation is $MgO + HNO_3 \rightarrow Mg(NO_3)_2 + H_2O$

Note that the nitrate group has not changed so count it as a separate group.

Count the magnesium atoms: 1 on the left; 1 on the right.

Count the oxygen atoms: 1 on the left; 1 on the right (remember, you are counting the nitrate group as a separate group, so do not count the oxygen atoms in this group).

Count the hydrogen atoms: 1 on the left; 2 on the right. Therefore you must change $HNO_3$ to $2HNO_3$.

The equation now reads $MgO + 2HNO_3 \rightarrow Mg(NO_3)_2 + H_2O$

Count the nitrate groups: 2 on the left; 2 on the right.

The equation is now balanced.

The final equation is: $MgO + 2HNO_3 \rightarrow Mg(NO_3)_2 + H_2O$

## Using state symbols in equations

State symbols are sometimes written after formulae in chemical equations to show which physical state each substance is in.

There are four different state symbols that you need to know. They are:

(s) – solid (l) – liquid (g) – gas (aq) – aqueous

An example of an equation with state symbols is

$$Zn(s) + CuSO_4(aq) \rightarrow ZnSO_4(aq) + Cu(s)$$

This equation tells you that when **solid** zinc is added to an **aqueous** solution of copper(II) sulfate, an **aqueous** solution of zinc sulfate and **solid** copper are formed.

State symbols are very useful, since they tell you the **conditions** required for a reaction to take place. If they were left out in the above equation, then you might think that the reaction would take place if you added solid zinc to solid copper(II) sulfate. **This is not the case**; no reaction takes place under these conditions.

## Calculating empirical formulae

The **empirical formula** of a compound gives the **simplest whole-number ratio** of atoms of each element in the compound. It can be calculated from knowledge of the ratio of masses of each element in the compound.

For example, a compound that contains 10 g of hydrogen and 80 g of oxygen has an empirical formula of $H_2O$. This can be shown by the following calculation:

**Amount of hydrogen atoms = mass in grams ÷ $A_r$ of hydrogen = (10 ÷ 1) = 10 mol**

**Amount of oxygen atoms = mass in grams ÷ $A_r$ of oxygen = (80 ÷ 16) = 5 mol**

Therefore, the ratio of moles of hydrogen atoms to moles of oxygen atoms is 10:5.

This is 2:1 in its simplest form.

Since equal numbers of moles of atoms contain the same number of atoms, it follows that the ratio of hydrogen atoms to oxygen atoms is 2:1.

Hence the empirical formula is $H_2O$.

> The ratio of masses can also be given as a percentage. In this case 11.1% of the mass is hydrogen and 88.9% of the mass is oxygen. The calculation is performed in exactly the same way, dividing the percentages by the relative atomic masses.

### Worked Example 1

Percentage composition: Carbon = 92.31%; Hydrogen = 7.69%

| | C | H |
|---|---|---|
| **Divide by $A_r$ to obtain ratio of moles of atoms** | (92.31 ÷ 12 ) = 7.69 | (7.69 ÷ 1) = 7.69 |
| **Divide by smallest to find simplest whole-number ratio** | (7.69 ÷ 7.69) = 1 | (7.69 ÷ 7.69) = 1 |

**The empirical formula is CH**

### Worked Example 2

Percentage composition: Carbon = 27.27%; Oxygen = 72.73%

| | C | O |
|---|---|---|
| **Divide by $A_r$ to obtain ratio of moles of atoms** | (27.27 ÷ 12) = 2.27 | (72.73 ÷ 16) = 4.55 |
| **Divide by smallest to find simplest whole-number ratio** | (2.27 ÷ 2.27) = 1 | (4.55 ÷ 2.27) ≈ 2 |

**The empirical formula is $CO_2$**

### Worked Example 3

Percentage composition: Aluminium = 52.94%; Oxygen = 47.06%

| | *Al* | O |
|---|---|---|
| **Divide by $A_r$ to obtain ratio of moles of atoms** | (52.94 ÷ 27) = 1.96 | (47.06 ÷ 16) = 2.94 |
| **Divide by smallest to find simplest whole-number ratio** | (1.96 ÷ 1.96) = 1 | (2.94 ÷ 1.96) = 1.5 |

Simplest whole-number ratio of 1 to 1.5 is 2 to 3.

**The empirical formula is $Al_2O_3$**

### Worked Example 4

Percentage composition: Copper = 37.43%; Chlorine = 41.52%; Water = 21.05%

| | Cu | *Cl* | $H_2O$ |
|---|---|---|---|
| **Divide by $A_r/M_r$ to obtain ratio of moles of atoms** | (37.43 ÷ 64) = 0.585 | (41.52 ÷ 35.5) = 1.17 | (21.05 ÷ 18) = 1.17 |
| **Divide by smallest to find simplest whole-number ratio** | (0.585 ÷ 0.585) = 1 | (1.17 ÷ 0.585) = 2 | (1.17 ÷ 0.585) = 2 |

**The empirical formula is $CuCl_2.2H_2O$**

## Calculating molecular formulae

The **molecular formula** gives the **exact** numbers of atoms of each element present in the formula of the compound.

Relationship between empirical and molecular formula:

| Name of compound | Empirical formula | Molecular formula |
|---|---|---|
| Methane | $CH_4$ | $CH_4$ |
| Ethane | $CH_3$ | $C_2H_6$ |
| Ethene | $CH_2$ | $C_2H_4$ |
| Benzene | CH | $C_6H_6$ |

If the empirical formula of a compound is known, then it is a simple matter to find its molecular formula as long as the $M_r$ of the compound is known. For example, the empirical formula of benzene is CH and its $M_r$ is 78.

The empirical formula mass is $(12 + 1) = 13$

$M_r$ divided by empirical formula mass is $78 \div 13 = 6$

**Therefore the molecular formula is six times (× 6) the empirical formula, i.e. $C_6H_6$**

**EXAMINER'S TIP**

The most common mistake here is to put a 6 in front of the empirical formula and give the answer as 6CH. This is **not** correct.

## Calculations using equations

### 1. Reacting masses

**Worked Example 1**

Calculate the mass of magnesium oxide that can be made by completely burning 6 g of magnesium in oxygen.

The equation for the reaction is: $2Mg + O_2 \rightarrow 2MgO$

There are two ways of solving this problem.

**Method 1 – Using moles**

***Step 1.*** Calculate the amount, in moles, of magnesium reacted

$A_r$ of Mg is 24

Amount of magnesium $= (6 \div 24) = 0.25$ mol

***Step 2.*** Calculate the amount of magnesium oxide formed

The equation tells us that 2 mol of Mg form 2 mol of MgO, hence the amount of MgO formed is the same as the amount of Mg reacted.

∴ Amount of magnesium oxide formed is 0.25 mol.

***Step 3.*** Calculate the mass of magnesium oxide formed

$M_r$ of MgO $+ (24 + 16) = 40$

**Mass of magnesium oxide $= (0.25 \times 40) = 10$ g**

**Method 2 – Using mass ratios**

$2Mg + O_2 \rightarrow 2MgO$

$A_r$ of Mg is 24; $M_r$ of MgO is 40

***Step 1.*** $2 \times 24$ g $= 48$ g of Mg $\rightarrow 2 \times 40$ g $= 80$ g of MgO

***Step 2.*** ∴ 6 g of Mg $\rightarrow (6 \div 48) \times 80$ g $=$ **10 g of MgO**

**EXAMINER'S TIP**

Using mass ratios is the easier method, particularly when the units of mass are not grams but kilograms or tonnes, since no conversion into grams is required to calculate the amount in moles. However, be careful, since you may not be given the choice; the question may be structured to make you use Method 1.

### Worked Example 2

Calculate the mass, in tonnes, of aluminium that can be produced from 51 tonnes of aluminium oxide.

The equation for the reaction is: $Al_2O_3 \rightarrow 2Al + 3O_2$

$A_r$ of $Al$ is 27 and $A_r$ of O is 16; ∴ $M_r$ of $Al_2O_3$ is 102

1 tonne $= 10^6$ g

**Method 1**

Amount of aluminium oxide reacted $= (51 \times 10^6) \div 102 = 500\,000$ mol

1 mol of $Al_2O_3$ produces 2 mol of $Al$

∴ Amount of $Al$ produced is $2 \times 500\,000$ mol $= 1\,000\,000$ mol

Mass of $Al$ produced $= (1\,000\,000 \times 27)$ g $= 27\,000\,000$ g

$27\,000\,000$ g $= (27\,000\,000 \div 10^6)$ tonnes $= 27$ tonnes

∴ **Mass of aluminium produced is 27 tonnes**

**Method 2**

102 tonnes of $Al_2O_3 \rightarrow (2 \times 27) = 54$ tonnes of $Al$

∴ 51 tonnes of $Al_2O_3 \rightarrow (51 \div 102) \times 54 =$ **27 tonnes of Al**

**EXAMINER'S TIP**

Notice how Method 2 avoids both the use of moles and the need to convert tonnes into grams and vice versa. If you are given the freedom to choose which method to use then you can use either.

## 2. Volumes of gases

The mole concept also allows us to work out the volumes of gases that react.

### Worked Example 1

Calculate the volume of gas (carbon dioxide) produced when 50 g of calcium carbonate is decomposed by heating.

***Step 1.*** Calculate the amount, in moles, of calcium carbonate reacted

$M_r$ of $CaCO_3$ is 100

Amount of $CaCO_3 = (50 \div 100) = 0.5$ mol

***Step 2.*** Calculate the amount, in moles, of carbon dioxide formed

$CaCO_3 \rightarrow CaO + CO_2$

1 mol of $CaCO_3$ produces 1 mol of $CO_2$

∴ 0.5 mol of $CaCO_3$ produces 0.5 mol of $CO_2$

***Step 3.*** Calculate the volume of $CO_2$ formed

Volume of $CO_2 = (0.5 \times 24)$ dm³ $=$ **12 dm³**

## Worked Example 2

Calculate the volume of hydrogen required to reduce 5 tonnes of copper(II) oxide to copper.

***Step 1.*** Calculate the amount, in moles, of copper(II) oxide reacted

$M_r$ of CuO is 80.

1 tonne $= 10^6$ g

Amount of CuO reacted $= (5 \times 10^6) \div 80 = 62\,500$ mol

***Step 2.*** Calculate the amount, in moles, of hydrogen required

$CuO + H_2 \rightarrow Cu + H_2O$

1 mol of CuO requires 1 mol of $H_2$

∴ 62 500 mol of CuO requires 62 500 mol of $H_2$

***Step 3.*** Calculate the volume of $H_2$ required

Volume of $H_2 = (62\,500 \times 24) =$ **1 500 000 dm³**

It is necessary to convert the mass from tonnes to grams before calculating the amount.

# Calculations involving concentrations of solutions

$$\textbf{Amount of dissolved substance (in mol)} = \frac{\textbf{volume of solution (in cm}^3\textbf{)} \times \textbf{concentration of solution (in mol/dm}^3\textbf{)}}{\textbf{1000}}$$

## Worked Example 1

Calculate the volume of hydrochloric acid of concentration 1.0 mol/dm³ that is required to react completely with 2.5 g of calcium carbonate.

***Step 1.*** Calculate the amount, in moles, of calcium carbonate that reacts

$M_r$ of $CaCO_3$ is 100

Amount of $CaCO_3 = (2.5 \div 100) = 0.025$ mol

***Step 2.*** Calculate the amount of hydrochloric acid required

$CaCO_3 + 2HCl \rightarrow CaCl_2 + H_2O + CO_2$

1 mol of $CaCO_3$ requires 2 mol of HCl

∴ 0.025 mol of $CaCO_3$ requires 0.05 mol of HCl

***Step 3.*** Calculate the volume of HCl required

Volume $=$ (amount $\times$ 1000) $\div$ concentration

$= (0.05 \times 1000) \div 1.0$

$=$ **50 cm³**

**Worked Example 2**

25.0 $cm^3$ of 0.050 $mol/dm^3$ sodium carbonate were completely neutralised by 20.00 $cm^3$ of dilute hydrochloric acid. Calculate the concentration, in $mol/dm^3$, of the hydrochloric acid.

***Step 1.*** Calculate the amount, in moles, of sodium carbonate reacted

$$\text{Amount of } Na_2CO_3 = \left(\frac{25.0 \times 0.050}{1000}\right) = 0.00125 \text{ mol}$$

***Step 2.*** Calculate the amount, in moles, of hydrochloric acid reacted

$$Na_2CO_3 + 2HCl \rightarrow 2NaCl + H_2O + CO_2$$

1 mol of $Na_2CO_3$ reacts with 2 mol of $HCl$

0.00125 mol of $Na_2CO_3$ reacts with 0.00250 mol of $HCl$

***Step 3.*** Calculate the concentration, in $mol/dm^3$, of the hydrochloric acid

$$\text{Concentration } (mol/dm^3) = \frac{(1000 \times \text{amount})}{\text{volume of solution}}$$

$$= \frac{(1000 \times 0.00250)}{20.0}$$

$$= \mathbf{0.125\ mol/dm^3}$$

**EXAMINER'S TIP**

The final answer is given to 3 significant figures, since the minimum number of significant figures supplied in the data of the question is 3. However, in this case, it is highly likely that an answer of 0.13 $mol/dm^3$ would be marked correct.

**Worked Example 3**

Calculate the volume, in $cm^3$, of carbon dioxide produced when an excess of sodium carbonate is reacted with 25 $cm^3$ of 2.0 $mol/dm^3$ sulfuric acid.

***Step 1.*** Calculate the amount, in moles, of sulfuric acid reacted

$$\text{Amount of } H_2SO_4 = \frac{25.0 \times 2.0}{1000} = 0.050 \text{ mol}$$

***Step 2.*** Calculate the amount of carbon dioxide that will be produced

$$Na_2CO_3 + H_2SO_4 \rightarrow Na_2SO_4 + H_2O + CO_2$$

1 mol of $H_2SO_4$ produces 1 mol of $CO_2$

0.050 mol of $H_2SO_4$ produces 0.050 mol of $CO_2$

***Step 3.*** Calculate the volume of carbon dioxide produced

$$\text{Volume of } CO_2 = 0.050 \times 24\,000 \text{ cm}^3$$

$$= \mathbf{1200\ cm^3}$$

Since the sodium carbonate is in excess, **all** of the sulfuric acid will react.

**EXAMINER'S TIP**

If you make a silly slip and calculate the volume in $dm^3$ instead of $cm^3$, as you are asked to do, you are unlikely to be penalised as long as you make it clear that your final answer is in $dm^3$. Hence, a final answer of 1.2 $dm^3$ would gain full marks.

## Calculations of percentage yield

Often in a chemical reaction the **theoretical yield** (the maximum amount of a product that could be formed from a given amount of a reactant) is not obtained. Under these circumstances it is useful to calculate the percentage yield to find out how efficient your reaction has been.

The percentage yield is calculated as follows:

$$\textbf{Percentage yield} = \frac{\textbf{yield obtained}}{\textbf{theoretical yield}} \times 100$$

**Worked Example**

In an experiment to displace copper from copper sulfate, 6.5 g of zinc was added to an excess of copper(II) sulfate solution. The copper was filtered off, washed and dried. The mass of copper obtained was 4.8 g. Calculate the percentage yield of copper.

The equation for the reaction is $Zn(s) + CuSO_4(aq) \rightarrow ZnSO_4(aq) + Cu(s)$

***Step 1.*** Calculate the amount, in moles, of zinc reacted

Amount of zinc $= \frac{6.5}{65} = 0.10\,mol$

***Step 2.*** Calculate the maximum amount of copper that could be formed

Maximum amount of copper $= 0.10\,mol$

***Step 3*** Calculate the maximum mass of copper that could be formed

Maximum mass of copper $= (0.10 \times 64) = 6.4\,g$

***Step 4.*** Calculate the percentage yield of copper

Percentage yield $= \frac{4.8}{6.4} \times 100 = 75\%$

Note that since the copper(II) sulfate is in excess, all of the zinc added would be expected to react.

# Chapter 6: Ionic compounds

## The formation of ions

An **ion** is an electrically charged atom or group of atoms. Ions are formed by the **loss** or **gain** of electrons.

The electronic configurations of the noble gases are particularly resistant to change; they are very reluctant to either lose or gain electrons. **Some** atoms of elements in Groups 1, 2, 3, 5, 6 and 7 of the Periodic Table form ions by the loss or gain of the number of electrons that results in the atoms obtaining the electronic configuration of their nearest noble gas (nearest, that is, in terms of atomic number, not necessarily distance apart in the Periodic Table).

The following examples will help you to understand the principles involved in working out the charges on the ions formed by some atoms.

**Worked Example 1**

What is the charge on the ion formed by an atom of lithium?

The electronic configuration of lithium is 2.1.

The nearest noble gas is helium, whose electronic configuration is 2.

A lithium atom therefore **loses one** electron.

Li 2.1 $\rightarrow$ $Li^+$ 2

The ion formed has three protons but only two electrons, hence there are three positive charges in the nucleus but only two negative charges outside of the nucleus. Therefore the ion has a net positive charge of 1 ($+3 - 2 = +1$). We represent the lithium ion by the formula $Li^+$.

### Worked Example 2

What is the charge on the ion formed by an atom of fluorine?

The electronic configuration of fluorine is 2.7.

The nearest noble gas is neon, whose electronic configuration is 2.8.

A fluorine atom therefore **gains one** electron.

F 2.7 $\rightarrow$ $F^-$ 2.8

The ion formed by fluorine has nine protons and ten electrons. It will therefore have a net negative charge of 1 ($+9 - 10 = -1$). The formula of the ion formed (which is called the **fluoride** ion) is $F^-$.

### Worked Example 3

What are the charges on the ions formed by magnesium and oxygen?

Similarly, the formation of magnesium and oxide ions can be represented by:

Mg 2.8.2 $\rightarrow$ $Mg^{2+}$ 2.8

magnesium ion (12 protons and 10 electrons)

O 2.6 $\rightarrow$ $O^{2-}$ 2.8

oxide ion (8 protons and 10 electrons)

The formulae of some common ions are shown in the tables below.

| Positive ions | |
|---|---|
| **Name of ion** | **Formula** |
| Lithium | $Li^+$ |
| Sodium | $Na^+$ |
| Potassium | $K^+$ |
| Silver | $Ag^+$ |
| Copper(I) | $Cu^+$ |
| Ammonium | $NH_4^+$ |
| Magnesium | $Mg^{2+}$ |
| Calcium | $Ca^{2+}$ |
| Barium | $Ba^{2+}$ |
| Zinc | $Zn^{2+}$ |
| Copper(II) | $Cu^{2+}$ |
| Lead(II) | $Pb^{2+}$ |
| Iron(II) | $Fe^{2+}$ |
| Aluminium | $Al^{3+}$ |
| Chromium(III) | $Cr^{3+}$ |
| Iron(III) | $Fe^{3+}$ |
| Lead(IV) | $Pb^{4+}$ |
| Manganese(IV) | $Mn^{4+}$ |

| Negative ions | |
|---|---|
| **Name of ion** | **Formula** |
| Fluoride | $F^-$ |
| Chloride | $Cl^-$ |
| Bromide | $Br^-$ |
| Iodide | $I^-$ |
| Hydroxide | $OH^-$ |
| Nitrate | $NO_3^-$ |
| Oxide | $O^{2-}$ |
| Sulfide | $S^{2-}$ |
| Sulfate | $SO_4^{2-}$ |
| Carbonate | $CO_3^{2-}$ |

## Formation of ionic compounds

Atoms of metals often transfer electrons to atoms of non-metals to form compounds made up of ions. Such compounds are called **ionic** compounds.

Sodium chloride, $NaCl$, is a typical ionic compound.

Sodium atoms transfer electrons to chlorine atoms to form sodium ions ($Na^+$) and chloride ions ($Cl^-$).

Each sodium atom transfers one electron to each chlorine atom. This can be represented by either an **electronic configuration diagram:**

| | | |
|---|---|---|
| Na 2.8.1 | $\rightarrow$ | $Na^+$ 2.8 |
| $Cl$ 2.8.7 | | $Cl^-$ 2.8.8 |

Or a **dot-and-cross diagram:**

Another typical ionic compound is magnesium oxide, MgO.

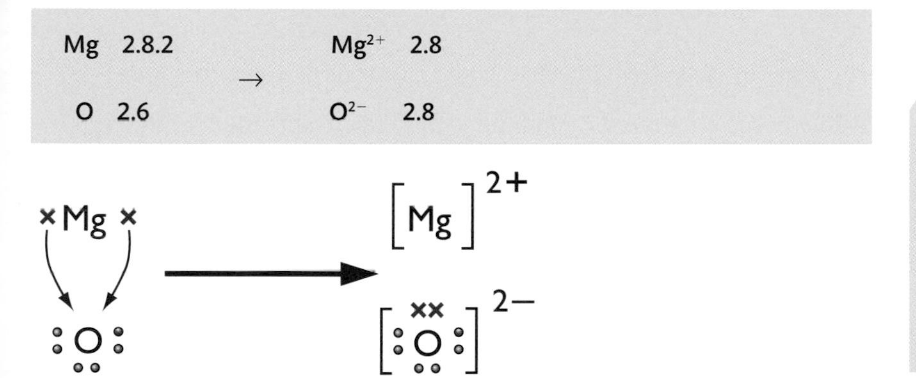


**EXAMINER'S TIP**

You need to be able to draw both electronic configuration diagrams and dot-and-cross diagrams for any of the metals from Groups 1, 2 and 3 combining with any of the non-metals from Groups 5, 6 and 7.

Each magnesium atom transfers two electrons to each oxygen atom.

When an ionic compound forms, the positively charged ions attract the negatively charged ions and arrange themselves into a three-dimensional structure called an **ionic lattice**. The arrangement of the ions in the ionic lattice of sodium chloride is shown below.

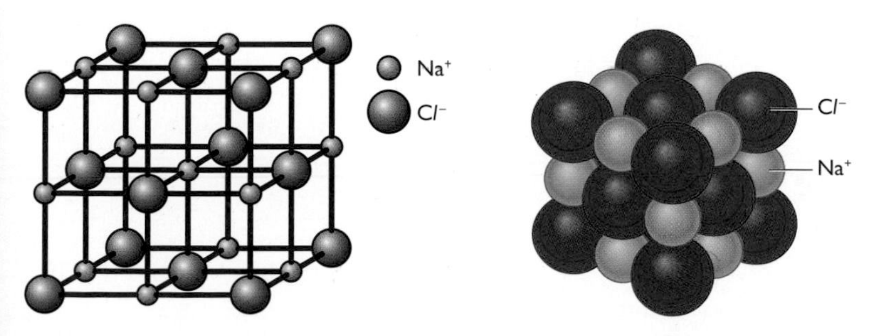


**Figure 6.1** *A diagram of the ionic lattice arrangement of sodium chloride and a model of a small part of a sodium chloride crystal.*

**Warning!** The lines of this diagram are **not** covalent bonds. They are just there to help show the arrangement of the ions. Those ions joined by lines are touching each other, as shown in the image of the model.

The structure of an ionic compound such as sodium chloride is described as **giant ionic**. There are **no molecules** in an ionic compound.

The electrostatic force of attraction between the oppositely charged ions is called an **ionic bond**.

The electrostatic forces are strong and there are many of them to break in an ionic crystal. A large amount of energy is, therefore, required to overcome the forces. For this reason, ionic compounds have high melting and boiling points.

For ions of similar size, the strength of the forces of attraction between the ions will depend on the size of their charge. For this reason, magnesium oxide ($Mg^{2+}O^{2-}$) has a higher melting and boiling point than sodium chloride ($Na^{+}Cl^{-}$). The 2+ and 2− ions in MgO attract one another more strongly than the 1+ and 1− ions in NaCl.

The melting point of aluminium oxide, $(Al^{3+})_2(O^{2-})_3$, is so high that it is difficult to find a substance with a high enough melting point to hold molten aluminium oxide. This is one of the reasons why aluminium oxide is dissolved in molten cryolite in the manufacture of aluminium *(see Chapter 25)*.

This table shows the melting and boiling points of sodium chloride and magnesium oxide.

| | Melting point in °C | Boiling point in °C |
|---|---|---|
| sodium chloride | 801 | 2852 |
| magnesium oxide | 1413 | 3600 |

# Chapter 7: Covalent substances

When atoms of non-metallic elements combine together they often share electrons between them.

When an atom of hydrogen, H, and an atom of chlorine, Cl, combine to form hydrogen chloride, HCl, the atoms come close enough together for their outer electron (valence) shells to overlap. This can be represented by the diagram below:

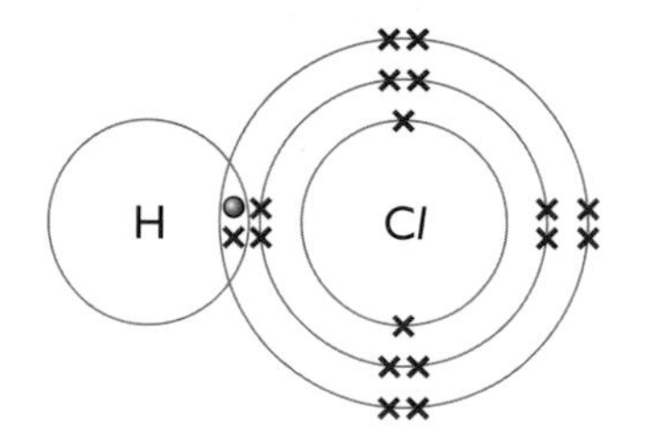


In this way, the shared region of the electron shells bonds the two nuclei together.

In the above example of hydrogen chloride, the shared region of electron shells contains two electrons; one electron was supplied by the hydrogen atom, the other by the chlorine atom. A bond formed in this way is called a **covalent bond**. Hydrogen chloride is a **covalent compound**.

Since the new particle formed is a molecule, hydrogen chloride is described as being a **simple molecular compound**. Its structure is said to be **simple molecular** because it consists of individual molecules.

It is only necessary to show the outer shell (valence) electrons.

The bonding in hydrogen chloride can be represented by a dot-and-cross diagram.

**EXAMINER'S TIP**

There is no need to draw the circles representing the outer shells when you are drawing dot-and-cross diagrams.

## Simple molecular substances

Both elements and compounds can exist as simple molecular substances.

The elements hydrogen, fluorine, chlorine, bromine, iodine, oxygen and nitrogen all exist as **diatomic molecules** under normal conditions. ('Diatomic' means that the molecule contains **two** atoms.)

The dot-and-cross diagrams for these molecules are as shown below:

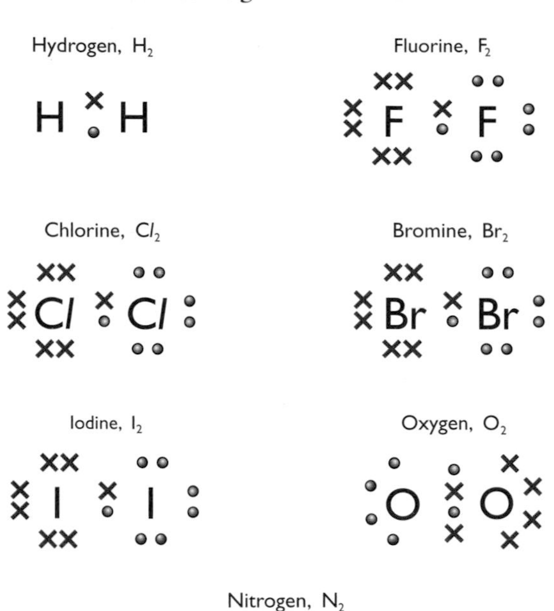


Notice that the dot-and-cross diagrams for the $F_2$, $Cl_2$, $Br_2$ and $I_2$ molecules are identical. This is because they belong to the same group of elements, the halogens – Group 7. Each halogen atom has seven electrons in its outer (valence) shell.

Also notice that the two atoms in the oxygen molecule share **four** electrons between them. Two electrons are supplied by one of the oxygen atoms and two are supplied by the other oxygen atom. A bond that is formed this way is called a **double** bond.

The bond between the two nitrogen atoms is a **triple** bond, since **six** electrons are shared between the two atoms. The table below shows the name, molecular formula and dot-and-cross diagram for six simple molecular compounds:

| Name of compound | Molecular formula | Dot-and-cross diagram |
| --- | --- | --- |
| Methane | $CH_4$ | |
| Ammonia | $NH_3$ | |
| Water | $H_2O$ | |
| Carbon dioxide | $CO_2$ | |
| Ethane | $C_2H_6$ | |
| Ethene | $C_2H_4$ | |

## Displayed formulae

Sometimes it is not necessary to show the bonding in covalent substances by using dots and crosses.

An alternative is to use an unbroken line to represent a shared pair of electrons, i.e. a single covalent bond.

A double covalent bond is indicated by using a double line ( = ).

Similarly, a triple bond is indicated by using a triple line ( ≡ ).

Diagrams drawn with lines to represent covalent bonds are called **displayed formulae**.

The displayed formulae of some of the simple molecular substances above are shown in the following table:

| Molecular formula | Displayed formula |
|---|---|
| $H_2$ | H−H |
| $Cl_2$ | Cl−Cl |
| $O_2$ | O=O |
| $N_2$ | N≡N |

| Molecular formula | Displayed formula |
|---|---|
| $CH_4$ | H<br>\|<br>H−C−H<br>\|<br>H |
| $H_2O$ | H−O−H |
| $C_2H_6$ | H H<br>\| \|<br>H−C−C−H<br>\| \|<br>H H |
| $C_2H_4$ | H H<br>\| \|<br>C=C<br>\| \|<br>H H |

Displayed formulae are particularly useful in Organic Chemistry ***(see Section C)***.

## Properties of simple molecular substances

Simple molecular substances usually have low melting and boiling points.

Iodine sublimes at normal atmospheric pressure, but it can be made to melt when put under higher pressures.

This is because the forces of attraction between the molecules are typically weak compared to the electrostatic forces of attraction between oppositely charged ions (i.e. ionic bonds) and therefore very little energy is required to overcome them.

Hence, simple molecular substances are often gases (e.g. hydrogen, oxygen, nitrogen and methane), or low boiling point liquids (e.g. water and bromine), or low melting point solids (e.g. iodine) at room temperature (25°C).

It is important to remember that when simple molecular substances change state, the covalent bonds between the atoms are **not** usually broken. Covalent bonds are strong compared to the forces of attraction between the molecules.

## Giant covalent structures

Some substances are made up of millions of atoms covalently bonded together to form a giant structure. Examples of giant covalent structures are diamond, graphite and silicon dioxide.

Diamond and graphite are two forms of the element carbon. The table shows the similarities and differences between them:

| | Diamond | Graphite |
|---|---|---|
| **Bonding** | Each carbon atom forms four single covalent bonds to other carbon atoms<br>A three-dimensional structure is formed<br>All four of the outer shell (valence) electrons are used to form covalent bonds | Each carbon atom forms three single covalent bonds to other carbon atoms<br>A layered structure is formed<br>Only three of the valence electrons are used to form covalent bonds. The fourth electron from each atom exists between the layers and is delocalised, as in the bonding in metals ***(see Chapter 8)*** |
| **Structure** | | one layer (top view)<br>how the layers fit together (side view) |
| **Forces** | The covalent bonds are strong. There are no weak forces in the structure | The covalent bonds are strong<br>The forces of attraction between the layers are weak |
| **Properties** | Very high melting point (many strong covalent bonds have to be broken, which requires a lot of heat energy)<br>Very hard and abrasive (strong covalent bonds are difficult to break)<br>Does not conduct electricity (all four valence electrons are used to form bonds; there are no electrons that are 'delocalised', or spread out - see Chapter 8 - and free to move) | Very high melting point (many strong covalent bonds have to be broken, which requires a lot of heat energy)<br>Soft and slippery (forces of attraction between the layers are weak so the layers easily slide over one another and can easily be separated)<br>Conducts electricity (only three valence electrons are used in forming covalent bonds; the fourth electron is delocalised between the layers and free to move parallel to the layers) |
| **Uses** | Cutting tools<br>Jewellery | Lubricant<br>Electrodes for electrolysis |

# Chapter 8: Metallic crystals

Metals have a giant, three-dimensional lattice structure in which positive ions are arranged in a regular pattern in a 'sea of electrons'. The outer shell (valence) electrons are detached from the atoms and are delocalised (spread out) throughout the structure.

electrons from outer shell of metal atoms

metal ions

**Figure 8.1** *Metallic crystal.*

The attraction between the positive ions and the delocalised electrons is known as a **metallic bond**, and this attraction keeps the ions together.

**EXAMINER'S TIP**

A common mistake is to refer to the particles in a metal as atoms or even protons or nuclei. Since the valence electrons are detached, the particles are positive ions, not atoms.

## Explaining the properties of metals

| Property | Reason |
|---|---|
| Most metals have high melting and boiling points | Metallic bonds are strong and there are many of them to overcome in a giant structure, hence a lot of heat energy is required |
| Good conductors of electricity | The delocalised electrons are free to move when a potential difference is applied across the metal |
| Malleable and ductile | The layers of positive ions can easily slide over one another and take up different positions. The delocalised electrons move with them so the metallic bonds are not broken |

# Chapter 9: Electrolysis

Ionic compounds do not conduct electricity when solid since the ions are not free to move.

However, ionic compounds do conduct electricity when **molten** or when in **aqueous solution** (i.e. when dissolved in water).

When an electric current is passed through a molten ionic compound, or through an aqueous solution of an ionic compound, **electrolysis** takes place.

Electrolysis is the decomposition (i.e. the **chemical** breakdown) of a substance by passing an electric current through it.

## Electrolysis of molten compounds

The apparatus shown below can be used to demonstrate the electrolysis of a molten ionic compound such as lead(II) bromide:

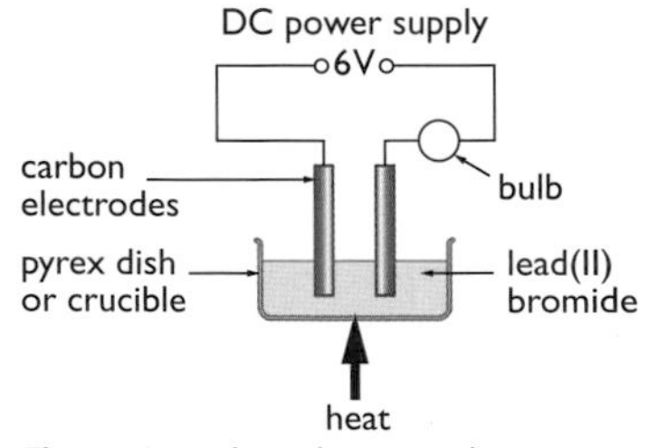


**Figure 9.1** *Electrolysing molten lead(II) bromide.*

The substance that is being electrolysed is called the **electrolyte**. The electrical connections between the electrolyte and the external electrical circuit are called **electrodes**. The positive electrode is called an **anode**. The negative electrode is called a **cathode**.

You do not need to remember the names of the two electrodes, but if you decide to use them in an examination then make sure you get them the correct way round.

The products of electrolysing a molten binary compound (i.e. a compound containing only **two** elements) can easily be predicted by using the following rules:

- The metal is formed at the negative electrode
- The non-metal is formed at the positive electrode

For example, with molten lead(II) bromide, lead is formed at the negative electrode and bromine is formed at the positive electrode.

Further examples of electrolysis of molten binary compounds are shown in the table below:

| Compound | Product at negative electrode | Product at positive electrode |
|---|---|---|
| Sodium chloride, $NaCl$ | Sodium, Na | Chlorine, $Cl_2$ |
| Magnesium fluoride, $MgF_2$ | Magnesium, Mg | Fluorine, $F_2$ |
| Aluminium oxide, $Al_2O_3$ | Aluminium, $Al$ | Oxygen, $O_2$ |

## How does electrolysis work?

### *Reaction at the negative electrode (cathode)*

Positive metal ions are attracted to the negative electrode. When they get to the electrode, they **gain** electrons so that they are converted into their **atoms**.

For example, when lead(II) bromide is electrolysed, lead(II) ions gain electrons to form lead atoms:

$$Pb^{2+} + 2e^- \rightarrow Pb$$

Lead forms as a liquid since the temperature of the molten electrolyte is above the melting point of lead.

⁂ Lead is formed at the negative electrode.

### *Reaction at the positive electrode (anode)*

Negative non-metal ions are attracted to the positive electrode. When they get to the electrode, they **lose** electrons to form atoms. For example, when lead(II) bromide is electrolysed, bromide ions lose an electron to form bromine atoms:

$$Br^- \rightarrow Br + e^-$$

Bromine forms as a gas since the temperature of the molten electrolyte is above the boiling point of bromine.

These atoms then pair up to form molecules:

$$Br + Br \rightarrow Br_2$$

The two equations above can be combined into one, **overall** equation:

$$2Br^- \rightarrow Br_2 + 2e^-$$

⁂ Bromine is formed at the positive electrode.

**EXAMINER'S TIP**
It is very likely that the equation you will be asked to produce in an examination for the reaction at the positive electrode is the **final, overall** equation.

For more electrode reactions, look at the table below:

| Compound | Reaction at negative electrode (cathode) | Reaction at positive electrode (anode) |
|---|---|---|
| Sodium chloride, $NaCl$ | $Na^+ + e^- \rightarrow Na$ | $2Cl^- \rightarrow Cl_2 + 2e^-$ |
| Magnesium fluoride, $MgF_2$ | $Mg^{2+} + 2e^- \rightarrow Mg$ | $2F^- \rightarrow F_2 + 2e^-$ |
| Aluminium oxide, $Al_2O_3$ | $Al^{3+} + 3e^- \rightarrow Al$ | $2O^{2-} \rightarrow O_2 + 4e^-$ |

## Electrolysis of aqueous solutions

Any aqueous solution that contains ions can be electrolysed. This includes aqueous solutions of ionic compounds such as the salts sodium chloride and copper(II) sulfate and also aqueous solutions of acids, such as sulfuric acid. The products are not as easy to predict as with molten compounds, since water is present and this has an influence on the reactions that take place at the electrodes.

The table below shows the products of electrolysing several aqueous solutions of salts and acids:

| Aqueous solution | Product at negative electrode | Product at positive electrode | Substance remaining in solution |
|---|---|---|---|
| sulfuric acid | hydrogen | oxygen | sulfuric acid (more concentrated) |
| nitric acid | hydrogen | oxygen | nitric acid (more concentrated) |
| sodium nitrate | hydrogen | oxygen | sodium nitrate (more concentrated) |
| sodium chloride | hydrogen | chlorine | sodium hydroxide |
| potassium sulfate | hydrogen | oxygen | potassium sulfate (more concentrated) |
| copper(II) sulfate | copper | oxygen | sulfuric acid |
| silver nitrate | silver | oxygen | nitric acid |

From the above results the following rules can be obtained:

**1.** The product at the **negative electrode** is either **hydrogen** or a **metal**.

**2.** The product at the **positive electrode** is either **oxygen** or **another non-metal** (such as chlorine).

*See Chapter 15 for details of the reactivity series.*

If the metal in the salt is **above** hydrogen in the reactivity series then **hydrogen** is evolved at the cathode.

If the metal in the salt is **below** hydrogen in the reactivity series then the **metal** is deposited on the cathode.

### Explaining the electrolysis of aqueous solutions

We will now look at the electrolysis of three aqueous solutions in more detail.

### *1. Electrolysis of aqueous sodium chloride using inert electrodes*

Suitable materials for the electrodes are platinum or carbon (graphite).

Particles present in solution:

1. Sodium ions, $Na^+$
2. Chloride ions, $Cl^-$
3. Water molecules, $H_2O$

**Reaction at the negative electrode:**

Hydrogen gas is produced as water molecules gain electrons:

$$2H_2O + 2e^- \rightarrow H_2 + 2OH^-$$

**Reaction at the positive electrode:**

Chlorine gas is produced as chloride ions lose electrons:

$$2Cl^- \rightarrow Cl_2 + 2e^-$$

**Overall reaction:**

$$2NaCl(aq) + 2H_2O(l) \rightarrow 2NaOH(aq) + H_2(g) + Cl_2(g)$$

The sodium chloride solution is gradually converted into sodium hydroxide solution.

**Note 1:** The solution around the negative electrode becomes alkaline owing to the increased concentration of hydroxide ions *(see Chapter 21)*.

**Note 2:** The formation of hydrogen molecules and hydroxide ions is sometimes explained by the following two reactions.

**A.** Water molecules self-ionise to produce hydrogen ions and hydroxide ions:

$$H_2O \rightarrow H^+ + OH^-$$

**B.** The hydrogen ions from the self-ionisation of water molecules are discharged to form hydrogen molecules, leaving behind hydroxide ions in solution:

$$2H^+ + 2e^- \rightarrow H_2$$

This second part of the explanation is not very satisfactory since the number of hydrogen ions in a **neutral solution** such as aqueous sodium chloride is considerably lower than the number of water molecules.

## *2. Electrolysis of dilute sulfuric acid using inert electrodes*

A suitable material for the electrodes is platinum. Carbon (graphite) is not suitable since it will react with the oxygen produced at the positive electrode.

Particles present in solution:

1. Hydrogen ions, $H^+$
2. Sulfate ions, $SO_4^{2-}$
3. Water molecules, $H_2O$

**Reaction at the negative electrode:**

Hydrogen gas is produced as hydrogen ions gain electrons:

$$2H^+ + 2e^- \rightarrow H_2$$

The formation of hydrogen can also be explained by the water molecules gaining electrons. This time, the solution around the negative electrode does not become alkaline as the hydroxide ions formed are neutralised by the hydrogen ions from the sulfuric acid to form water molecules *(see Chapter 21)*.

**Reaction at the positive electrode:**

Oxygen gas is produced as water molecules lose electrons:

$$2H_2O \rightarrow O_2 + 4H^+ + 4e^-$$

**Overall reaction:**

$$2H_2O(l) \rightarrow 2H_2(g) + O_2(g)$$

As the water is decomposed, the solution of sulfuric acid gradually becomes more concentrated.

The formation of oxygen molecules is sometimes explained by the following two reactions.

**A.** Water molecules self-ionise to form hydrogen ions and hydroxide ions:

$$H_2O \rightarrow H^+ + OH^-$$

**B.** The hydroxide ions are discharged to form oxygen molecules:

$$4OH^- \rightarrow O_2 + 2H_2O + 4e^-$$

This second part of the explanation is not very satisfactory since the concentration of hydroxide ions in the **strongly acidic** solution of sulfuric acid is very low.

### *3. Electrolysis of aqueous copper(II) sulfate using inert electrodes*

A suitable material for the electrodes is platinum. Carbon (graphite) is not suitable since it will react with the oxygen produced at the positive electrode.

Particles present in solution:
1. Copper(II) ions, $Cu^{2+}$
2. Sulfate ions, $SO_4^{2-}$
3. Water molecules, $H_2O$

**Reaction at the negative electrode:**

Copper is deposited on the negative electrode as copper(II) ions gain electrons:

$$Cu^{2+} + 2e^- \rightarrow Cu$$

**Note 1:** The blue colour of the solution gradually fades, and eventually becomes colourless, as the copper(II) ions are removed from solution.

**Note 2:** The copper(II) sulfate solution gradually changes into sulfuric acid, since the copper(II) ions are replaced by hydrogen ions.

**Reaction at the positive electrode:**

Oxygen gas is produced as water molecules lose electrons:

$$2H_2O \rightarrow O_2 + 4H^+ + 4e^-$$

**Overall reaction:**

$$2CuSO_4(aq) + 2H_2O(l) \rightarrow 2H_2SO_4(aq) + 2Cu(s) + O_2(g)$$

## Electrolysis calculations

When a current passes through a solution of a salt of a metal that is low in the reactivity series, metal ions are discharged. As a result, metal atoms are deposited on the negative electrode.

For silver ions, the equation for the reaction at the negative electrode is:

| $Ag^+(aq)$ | + | $e^-$ | $\rightarrow$ | Ag |
|---|---|---|---|---|
| 1 mol | | 1 mol | | 1 mol |

This equation tells us that one mole (1 mol) of silver ions accepts one mole (1 mol) of electrons to form one mole (1 mol) of silver atoms.

The quantity of electrical charge required to deposit 1 mol (i.e. 108 g) of silver is found by experiment to be 96 500 coulombs (96 500 C).

96 500 C is called one faraday (1 F) of electricity and represents one mole of electrons.

One coulomb is the quantity of electrical charge produced by the passage of one ampere (1A) for one second (1s).

Quantity of electrical charge, current and time are therefore linked by the following equation:

$$\text{Quantity (in coulombs)} = \text{Current (in amperes)} \times \text{Time (in seconds)}$$

or

$$Q = It$$

96 500 C must be the electrical charge on one mole (1 mol) of electrons (i.e. $6 \times 10^{23}$ electrons). The value 96 500 C/mol is called the **Faraday constant**, after the famous scientist Michael Faraday.

When copper is deposited in electrolysis, the electrode reaction is:

| $Cu^{2+}(aq)$ | + | $2e^-$ | → | $Cu(s)$ |
|---|---|---|---|---|
| 1 mol | | 2 mol | | 1 mol |

Thus, 1 mol of copper ions requires 2 mol of electrons, or $2 \times 96\,500$ C, for discharge.

When gold is deposited in electrolysis, the electrode reaction is:

| $Au^{3+}(aq)$ | + | $3e^-$ | → | $Au(s)$ |
|---|---|---|---|---|
| 1 mol | | 3 mol | | 1 mol |

Thus, 1 mol of gold ions requires 3 mol of electrons, or $3 \times 96\,500$ C, for discharge.

### Worked Example 1

Calculate the masses of metals deposited when one faraday (1 F) of electricity flows through each of the following solutions: (a) silver nitrate, (b) nickel(II) nitrate, and (c) aluminium sulfate.

Start with the equations: $Ag^+(aq) + e^- \rightarrow Ag$

$Ni^{2+}(aq) + 2e^- \rightarrow Ni(s)$

$Al^{3+}(aq) + 3e^- \rightarrow Al(s)$

One faraday (1 F) of electricity supplies one mole (1 mol) of electrons.

From the equations,

1 mol of electrons deposit 1 mol of silver = **108 g silver**

1 mol of electrons deposit $\frac{1}{2}$ mol of nickel = $\frac{1}{2} \times 59$ g = **29.5 g nickel**

1 mol of electrons deposit $\frac{1}{3}$ mol of aluminium = $\frac{1}{3} \times 27$ g = **9 g aluminium**

### Worked Example 2

A current of 0.010 amperes (0.010 A) passes for 4.00 hours through a solution of gold(III) nitrate. What mass of metal is deposited?

Quantity of electrical charge = $0.010 \times 4.00 \times 60 \times 60$ C

= 144 C

Equation: $Au^{3+}(aq) + 3e^- \rightarrow Au(s)$

3 mol → 1 mol

Therefore, $3 \times 96\,500$ C deposit 1 mol (197 g) gold

Hence, 144 C deposit $\frac{197}{3 \times 96\,500} \times 144$ g gold = **0.098 g gold**

**Worked Example 3**

A metal of relative atomic mass 48.0 is deposited by electrolysis. If 0.239 g of the metal is deposited when 0.100 A flow for 4.00 hours, what is the charge on the ion of this element?

Quantity of electrical charge $= 0.100 \times 4.00 \times 60 \times 60\,C$

$= 1440\,C$

0.239 g of metal is deposited by 1 440 C

$\therefore$ 48.0 g of metal is deposited by $\frac{1440}{0.239} \times 48.0 = 289\,205\,C$

$289\,205\,C = \frac{289\,205}{96\,500}\,F = 3\,F$ (i.e. 3 mol of electrons)

1 mol of metal atoms is deposited by 3 mol of electrons

**The charge on the metal ion is $3^+$.**

## Calculating the volume of gases produced during electrolysis

### *Hydrogen*

The equation for the production of hydrogen at the negative electrode may be either

**$2H_2O(l) + 2e^- \rightarrow H_2(g) + 2OH^-(aq)$, in neutral solutions of salts**

**or**

**$2H^+(aq) + 2e^- \rightarrow H_2(g)$ in acidic solutions (e.g. sulfuric acid)**

In either case, 1 mol of hydrogen molecules is produced from 2 mol of electrons.

This means that $24\,dm^3$ ($24\,000\,cm^3$) of hydrogen at r.t.p. is produced from $2 \times 96\,500\,C$ (2 F) of electricity.

The equation $4OH^-(aq) - 4e^- \rightarrow 2H_2O(l) + O_2(g)$ is sometimes used by textbooks to represent the production of oxygen at the anode in the electrolysis of aqueous solutions of salts, acids and alkalis. This equation also leads to the formation of 1 mol of oxygen molecules from 4 mol of electrons.

### *Oxygen*

The equation for the production of oxygen at the anode is

**$2H_2O(l) \rightarrow 4H^+(aq) + O_2(g) + 4e^-$**

4 mol of electrons produce 1 mol of oxygen molecules.

Hence, $4 \times 96\,500\,C$ (4 F) produce $24\,dm^3$ ($24\,000\,cm^3$) of oxygen at r.t.p.

### *Chlorine*

The equation for the production of chlorine at the anode is

**$2Cl^-(aq) \rightarrow Cl_2(g) + 2e^-$**

Therefore, 2 mol of electrons produce 1 mol of chlorine molecules.

Hence, $2 \times 96\,500\,C$ (2 F) produces $24\,dm^3$ ($24\,000\,cm^3$) of chlorine at r.t.p.

## Worked Example

***a)*** Calculate the volume of **(i)** hydrogen and **(ii)** oxygen, measured at r.t.p., that should be formed when a current of 10.72 mA is passed for 5.00 hours through a solution of sulfuric acid.

[1 mol of gas occupies 24 000 $cm^3$ at r.t.p.]

***b)*** In practice, the volume of oxygen collected at the positive electrode is less than that expected by calculation. Why is this?

***a)*** Quantity of electrical charge $= (10.72 \div 1000) \times 5.00 \times 60 \times 60\,C$

$= 193\,C$

$= 0.002\,F$

*(i)* $2H^+(aq) + 2e^- \rightarrow H_2(g)$

2 mol 1 mol

2 F of electrical charge produces 24 000 $cm^3$ of hydrogen

⁂ 0.002 F produces **24 $cm^3$ of hydrogen**

*(ii)* $H_2O(l) \rightarrow 4H^+(aq) + O_2(g) + 4e^-$

1 mol 4 mol

4 F of electrical charge produces 24 000 $cm^3$ of oxygen

⁂ 0.002 F produces **12 $cm^3$ of oxygen**

***b)*** Some of the oxygen formed dissolves in the water present in the dilute sulfuric acid.

## Examination Questions

**1** Atoms are made up of three types of particle: proton, neutron and electron.

***a)*** Which **one** of these particles has the smallest mass? *(1)*

***b)*** Which **one** of these particles has a negative charge? *(1)*

***c)*** Which **two** of these particles are present in the nucleus of an atom? *(1)*

***d)*** Which **two** of these particles are present in equal numbers in an atom? *(1)*

***e)*** Isotopes of an element have different numbers of **one** of these particles. Name this particle. *(1)*

*(Total 5 marks)*

**2** ***a)*** Some elements combine together to form ionic compounds. Use words from the box to complete the sentences.

Each word may be used once, more than once or not at all.

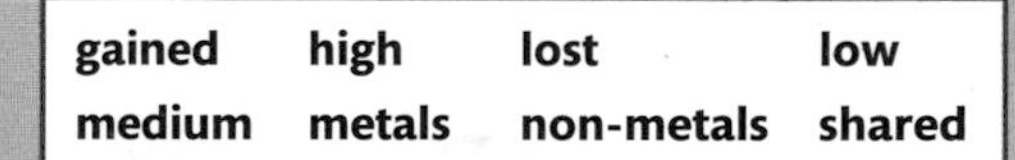

| **gained** | **high** | **lost** | **low** |
|---|---|---|---|
| **medium** | **metals** | **non-metals** | **shared** |

Ionic compounds are formed between ______________ and ______________.

Electrons are ______________ by atoms of one element and ______________ by atoms of the other element.

The ionic compound formed has a ______________ melting point and a ______________ boiling point. *(6)*

***b)*** Two elements react to form an ionic compound with the formula $MgCl_2$.

**(i)** Give the electronic configurations of the two elements in this compound **before** the reaction. *(2)*

**(ii)** Give the electronic configurations of the two elements in this compound **after** the reaction. *(2)*

*(Total 10 marks)*

**3** *a)* The diagrams show the arrangement of particles in the three states of matter. Each circle represents a particle.

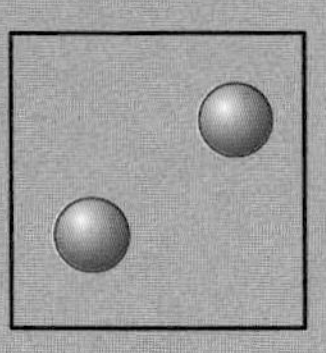
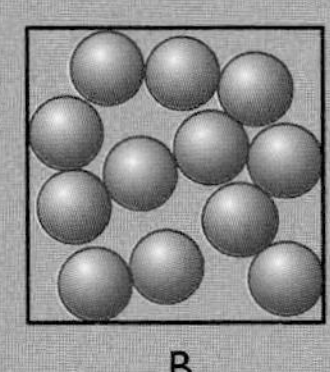
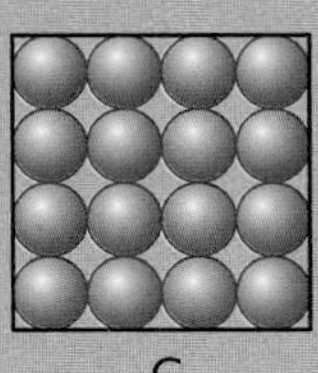

A B C

Use the letters **A**, **B** and **C** to give the starting and finishing states of matter of each of the changes in the table. *(4)*

| Change | Starting state | Finishing state |
|---|---|---|
| The formation of water vapour from a puddle of water on a hot day | | |
| The formation of solid iron from molten iron | | |
| The manufature of poly(ethene) from ethene | | |
| The reaction whose equation is ammonium chloride(s) → ammonia(g) + hydrogen chloride(g) | | |

*b)* Which state of matter is the **least** common for the elements of the Periodic Table at room temperature? *(1)*

*c)* The manufacture of sulfuric acid can be summarised by the equation:

$2S(s) + 3O_2(g) + 2H_2O(l) \rightarrow 2H_2SO_4(l)$

Tick one box in each line to show whether the formulae in the table below represent a compound, an element or a mixture. *(4)*

| | Compound | Element | Mixture |
|---|---|---|---|
| $2S(s)$ | | | |
| $2S(s) + 3O_2(g)$ | | | |
| $3O_2(g) + 2H_2O(l)$ | | | |
| $2H_2SO_4(l)$ | | | |

***(Total 9 marks)***

**4** A sample of the element rubidium, Rb, contains two isotopes.

*a)* Explain what isotopes are. *(2)*

*b)* **(i)** Complete the table for the isotopes of rubidium. *(3)*

| Atomic number of isotope | Mass number of isotope | Number of protons | Number of neutrons | Percentage of each isotope in sample |
|---|---|---|---|---|
| 37 | 85 | | | 72 |
| | | 37 | 50 | 28 |

**(ii)** Use the table to calculate the relative atomic mass of the sample of rubidium. Give your answer to one decimal place. *(2)*

*c)* Why do the two isotopes of rubidium have the same chemical properties? *(1)*

*d)* Rubidium reacts with oxygen, chlorine and water in a similar way to other elements in Group 1 of the Periodic Table.

**(i)** Suggest the formula of the compound formed when rubidium reacts with oxygen and chlorine. *(2)*

**(ii)** A small piece of rubidium is added to a trough of water. Suggest two observations you could make during the reaction. *(2)*

**(iii)** Complete and balance the equation for the reaction of rubidium with water.

$Rb + H_2O \rightarrow$ ______________ *(2)*

***(Total 14 marks)***

**5** The diagram shows the apparatus used to electrolyse lead(II) bromide.

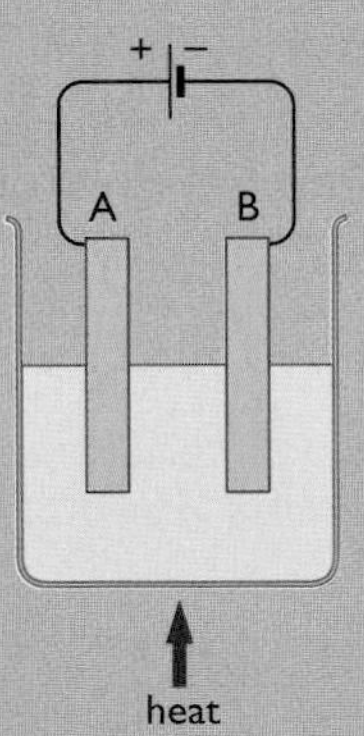


*a)* The wires connected to the electrodes are made of copper.

Explain why copper conducts electricity. *(1)*

*b)* Explain why electrolysis does not occur unless the lead(II) bromide is molten. *(2)*

*c)* The reactions occurring at the electrodes can be represented by the equations shown in the table.

Complete the table to show the electrode (**A** or **B**) at which each reaction occurs, and the type of reaction occurring (oxidation or reduction). *(2)*

| Electrode reaction | Electrode | Type of reaction |
|---|---|---|
| $Pb^{2+} + 2e^- \rightarrow Pb$ | | |
| $2Br^- \rightarrow Br_2 + 2e^-$ | | |

*d)* In an experiment using the same apparatus, the amount of charge passed was 0.10 faraday.

**(i)** Calculate the maximum amount, in moles, of Pb and the amount, in moles, of $Br_2$ formed. *(2)*

**(ii)** Calculate the mass of bromine formed. *(2)*

*(Total 9 marks)*

**6** *a)* A student made a solution of potassium hydroxide by dissolving 14.0 g of solid potassium hydroxide in distilled water to make 250 $cm^3$ of solution.

**(i)** Calculate the relative formula mass of potassium hydroxide, KOH. *(1)*

**(ii)** Calculate the amount, in moles, of potassium hydroxide in 14.0 g. *(1)*

**(iii)** Calculate the concentration, in mol/$dm^3$, of this solution of potassium hydroxide.

Show your working. *(2)*

*b)* A different solution of potassium hydroxide, of concentration 2.0 mol/$dm^3$, was used in an experiment to react with carbon dioxide gas.

The equation for this reaction is

$2KOH(aq) + CO_2(g) \rightarrow K_2CO_3(aq) + H_2O(l)$

**(i)** Calculate the amount, in moles, of potassium hydroxide in 200 $cm^3$ of this solution. *(1)*

**(ii)** Calculate the amount, in moles, of carbon dioxide that reacts with 200 $cm^3$ of this solution of potassium hydroxide. *(1)*

**(iii)** Calculate the volume that this amount of carbon dioxide occupies at room temperature and pressure (r.t.p.) *(1)*

(molar volume of any gas = 24 $dm^3$ at r.t.p.)

*(Total 7 marks)*

**7** Diamond and graphite are different forms of carbon.

*a)* State the term used to describe different forms of the same element in the same physical state. *(1)*

*b)* Name and describe the type of **bonding** in diamond. *(3)*

*c)* State **one** industrial use of diamond. *(1)*

*d)* Graphite has a hexagonal layer structure. Draw a diagram, showing three hexagons, to show the atoms and bonding in graphite. *(2)*

*e)* Diamond and graphite both have high melting points. Explain why. *(2)*

*(Total 9 marks)*

**8** This question is about two covalently bonded compounds.

*a)* The dot-and-cross diagram shows the covalent bonding in a hydrogen chloride molecule.

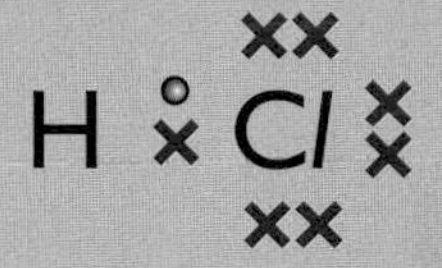


What is a covalent bond? *(1)*

*b)* Use words from the box to complete the sentences about hydrogen chloride. Each word may be used once, more than once or not at all.

| giant | high | ions | low |
|---|---|---|---|
| molecules | simple | strong | weak |

Hydrogen chloride has a ________________ molecular structure.

There are ________________ forces between the ________________. Because of this, hydrogen chloride has a ________________ boiling point. *(4)*

*c)* **(i)** Use the Periodic Table to help you complete the diagrams to show the electronic configuration of hydrogen and of oxygen. *(2)*

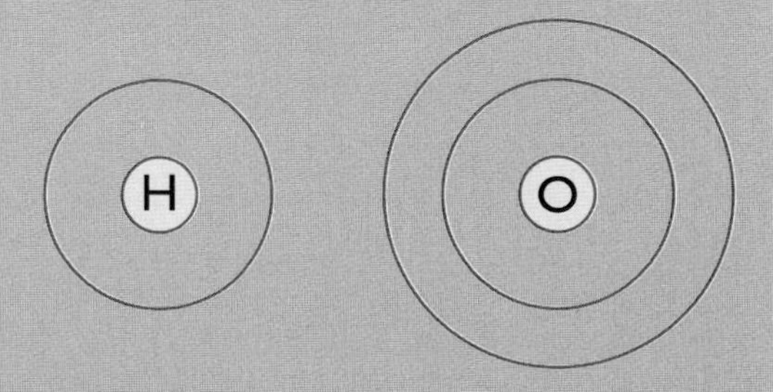


**(ii)** Draw a dot and cross diagram to show the covalent bonding in a water molecule. *(2)*

*(Total 9 marks)*

**9** The table gives the electronic configuration of three different atoms.

| Atom | Electronic configuration |
|---|---|
| fluorine | 2.7 |
| magnesium | 2.8.2 |
| sodium | 2.8.1 |

*a)* Describe the electron transfers that take place when magnesium reacts with fluorine to make the ionic compound magnesium fluoride, $MgF_2$. You may use diagrams to help your answer. *(3)*

*b)* In this reaction both oxidation and reduction have occurred. State which element has been oxidised, giving a reason. *(2)*

*c)* (i) Give the symbols of the ions formed by sodium and fluorine. *(1)*

(ii) Give the formula of sodium fluoride. *(1)*

*d)* A flame test is carried out on separate samples of magnesium fluoride and sodium fluoride.

The magnesium fluoride does not colour the flame.

What colour do you see when the sodium fluoride is tested? *(1)*

***(Total 8 marks)***

**10** In an experiment a student left some solid sodium chloride in a beaker of water for several days. The diagrams show the beaker at the start and end of the experiment.

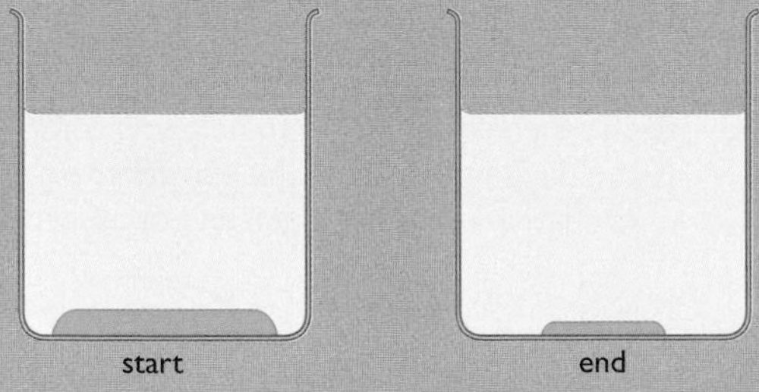


*a)* Write the formulae, with state symbols, of the two substances in the beaker at the start of the experiment. *(2)*

*b)* At the end of the experiment the student took a sample of the solution from near the top of the water. He tested it for the presence of chloride ions. The test was positive.

(i) Name the **two** substances the student added to test for the presence of chloride ions. *(2)*

(ii) Describe the observation made in the test. *(1)*

(iii) Name the process by which the chloride ions moved through the water to near the top of the water. *(1)*

*c)* Sea water contains dissolved sodium chloride. The following pieces of laboratory apparatus can be used to make drinking water from sea water.

(i) Draw a labelled diagram to show how these pieces of apparatus can be assembled to do this. *(3)*

(ii) Name the technique used in this process. *(1)*

***(Total 10 marks)***

**11** One reaction that occurs in the blast furnace during the extraction of iron is the reaction between iron(III) oxide and carbon:

$Fe_2O_3 + 3C \rightarrow 2Fe + 3CO$

*a)* Calculate the relative formula mass of iron(III) oxide, using information from the Periodic Table. *(1)*

*b)* 320 kg of iron(III) oxide were added to the blast furnace.

(i) Calculate the amount, in moles, of iron(III) oxide added. *(2)*

(ii) Calculate the maximum amount, in moles, of iron formed from this amount of iron(III) oxide. *(2)*

(iii) Calculate the maximum mass, in kilograms, of iron formed from this amount of iron(III) oxide. *(2)*

*c)* Most of the carbon monoxide formed in the reaction in ***b)*** is converted to carbon dioxide before it leaves the blast furnace.

(i) Explain how carbon monoxide acts as a poison. *(1)*

(ii) During one period in the operation of the blast furnace, the amount of carbon dioxide released was 5000 moles.

Calculate the volume, in $dm^3$, that this amount of carbon dioxide would occupy at room temperature and pressure (r.t.p.). *(1)*

(The molar volume of a gas is 24 $dm^3$ at r.t.p.)

*d)* Write the chemical equation for the reaction in which iron(III) oxide is reduced by carbon monoxide. *(2)*

*e)* **(i)** Limestone is added to the blast furnace to remove impurities. State the main impurity removed. *(1)*

**(ii)** Write **two** chemical equations to show how limestone removes this impurity. *(2)*

***(Total 14 marks)***

**12** The decomposition of ammonium chloride is a reversible reaction: $NH_4Cl(s) \rightleftharpoons NH_3(g) + HCl(g)$.

*a)* How is this reaction made to go in the **forward** direction? *(1)*

*b)* Concentrated hydrochloric acid gives off hydrogen chloride gas.

Concentrated ammonia solution gives off ammonia gas.

An experiment is set up.

cotton wool soaked in concentrated hydrochloric acid

glass tube

cotton wool soaked in concentrated ammonia solution

After a few minutes a white solid forms inside the tube. The solid forms when ammonia gas reacts wlth hydrogen chloride gas.

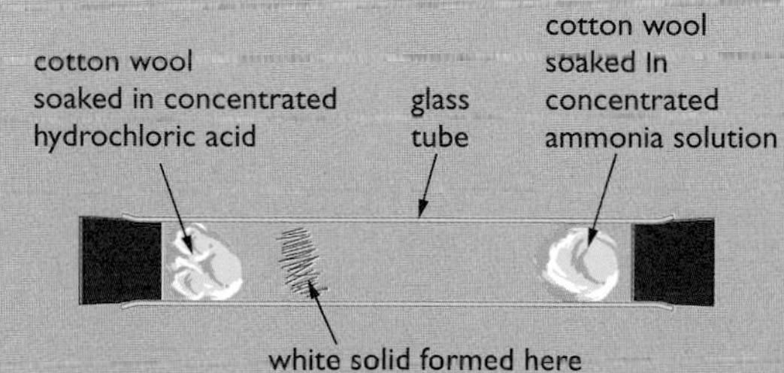


**(i)** Name the process by which the ammonia and hydrogen chloride particles move inside the tube. *(1)*

**(ii)** What is the white solid that forms inside the tube? *(1)*

**(iii)** What does the position of the white solid tell you about the relative speeds at which the ammonia and hydrogen chloride particles move? *(1)*

*c)* The experiment is repeated with a strip of damp red litmus paper placed along the inside of the tube.

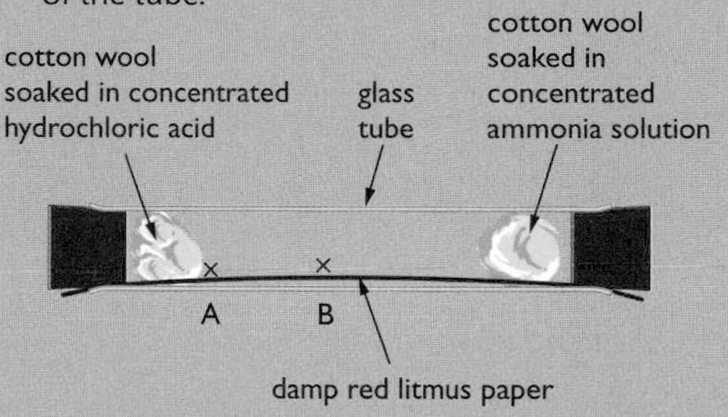


State the colour of the litmus paper at A and B when the white solid forms. *(2)*

***(Total 6 marks)***

**13** The method used to separate the substances in a mixture depends on the properties of the substances present.

For each of the following, name a suitable method for obtaining:

*a)* water from potassium chloride solution. *(1)*

*b)* potassium chloride from potassium chloride solution. *(1)*

*c)* water from a mixture of calcium carbonate and water. *(1)*

*d)* a red food dye from a mixture of coloured food dyes. *(1)*

*e)* gasoline from crude oil. *(1)*

***(Total 5 marks)***

**14** *a)* A solution was made by dissolving 1.62 g of hydrogen bromide, HBr, in 250 $cm^3$ of water.

**(i)** Calculate the relative formula mass of hydrogen bromide. Use data from the Periodic Table (p. 119). *(1)*

**(ii)** Calculate the amount, in moles, of hydrogen bromide in a 1.62 g sample. *(2)*

**(iii)** Calculate the concentration, in $mol/dm^3$, of the hydrogen bromide solution. *(2)*

**(iv)** Calculate the concentration, in $g/dm^3$, of the hydrogen bromide solution. *(2)*

**b)** Hydrogen bromide solution can be neutralised by adding sodium hydroxide solution.

A 20.0 $cm^3$ sample of a solution of hydrogen bromide had a concentration of 0.200 $mol/dm^3$.

(i) Write a chemical equation for this neutralisation reaction. *(1)*

(ii) Explain, with reference to protons, why this reaction is described as a neutralisation reaction. *(2)*

(iii) Calculate the amount, in moles, of hydrogen bromide in 20.0 $cm^3$ of 0.200 $mol/dm^3$ solution. *(2)*

(iv) Calculate the volume of 0.100 $mol/dm^3$ sodium hydroxide solution needed to neutralise this sample of hydrogen bromide solution. *(2)*

(v) Suggest the name of an indicator (other than litmus), and its colour change, that could be used to check when neutralisation was complete. *(3)*

***(Total 17 marks)***

**15** Magnesium and fluorine react together to form magnesium fluoride: $Mg(s) + F_2(g) \rightarrow MgF_2(s)$

**a)** (i) Describe the structure of a metal such as magnesium. *(2)*

(ii) What is meant by the term **malleable**? *(1)*

(iii) Explain, in terms of its structure, why magnesium is malleable. *(2)*

**b)** The atoms of fluorine in the $F_2$ molecule are joined by a covalent bond.

Describe how the atoms are held together by this bond. *(2)*

**c)** Give the electronic configuration of:

(i) a fluorine atom and (ii) a fluoride ion. *(2)*

**d)** Draw a diagram to show the arrangement of electrons in a magnesium ion, showing its charge. *(2)*

**e)** Suggest why magnesium fluoride, $MgF_2$, has a higher melting point than sodium fluoride, NaF. *(2)*

***(Total 13 marks)***

**16** The diagram shows apparatus that can be used to electrolyse dilute sulfuric acid.

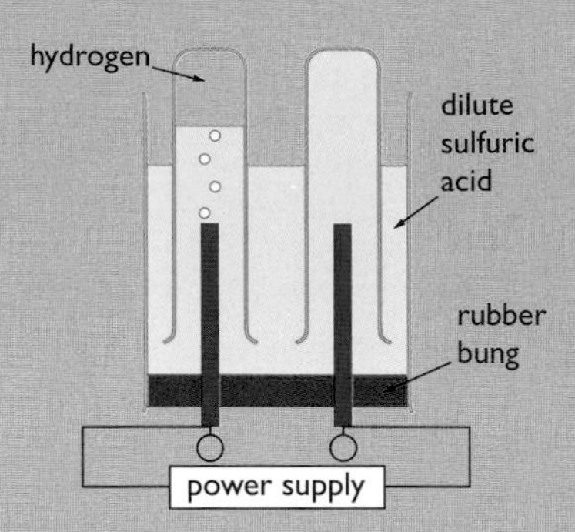


**a)** (i) Label the electrodes in the diagram by writing the symbols + and − in the circles. *(1)*

(ii) The equations for the reactions occurring at the electrodes are:

Equation **1**

$2H^+(aq) + 2e^- \rightarrow H_2(g)$

Equation **2**

$2H_2O(l) \rightarrow O_2(g) + 4H^+(aq) + 4e^-$

Give the formula of the species being reduced. Give a reason for your choice. *(2)*

(iii) The volume of hydrogen gas collected after a few minutes is shown on the diagram.

Draw another line on the diagram to show the volume of oxygen gas collected after the same length of time.

Explain your choice with reference to Equations **1** and **2**. *(3)*

**b)** In one experiment, the amount of charge passed was 0.40 faraday.

(i) Calculate the amount, in moles, of hydrogen gas formed. *(1)*

(ii) Calculate the volume, in $dm^3$, of this amount of hydrogen gas at room temperature and pressure (r.t.p.). *(2)*

(molar volume of any gas = 24 $dm^3$ at r.t.p.)

**c)** In a second experiment, the amount of charge passed was 0.80 faraday.

(i) Calculate the amount, in moles, of oxygen formed. *(1)*

(ii) Calculate the mass, in g, of oxygen formed. *(2)*

***(Total 12 marks)***

Notes

# Chapter 10: The Periodic Table

## Arranging the elements in order

First of all, some revision of earlier work:

- The atom is composed of protons, neutrons and electrons. The number of protons in an atom is the same as the number of electrons and is called the **atomic number**.
- Electrons are arranged in groups at different distances from the nucleus. The groups are called **shells**. The way the electrons are arranged in an atom is called its **electronic configuration**. The shell that is the furthest away from the nucleus is often referred to as the **outermost** shell (or simply the **outer** shell) or the **valence** shell.

Some interesting patterns emerge when the elements are placed in order of increasing atomic number and then arranged in horizontal rows in the form of a table – the Periodic Table. A new horizontal row is started after each noble gas.

Hydrogen and helium are often placed in Period 1 on their own.

After that, the next 18 elements are arranged as shown below to form three horizontal rows of elements labelled Period 2, Period 3 and Period 4 respectively.

In some Periodic Tables helium is placed at the top of Group 8. When this is done the group is usually labelled Group 0.

| Period | Group 1 | 2 | 3 | 4 | 5 | 6 | 7 | 8 |
|---|---|---|---|---|---|---|---|---|
| 1 | | H (1) | | He (2) | | | | |
| 2 | Li (2. 1) | Be (2. 2) | B (2. 3) | C (2. 4) | N (2. 5) | O (2. 6) | F (2. 7) | Ne (2. 8) |
| 3 | Na (2. 8. 1) | Mg (2. 8. 2) | Al (2. 8. 3) | Si (2. 8. 4) | P (2. 8. 5) | S (2. 8. 6) | Cl (2. 8. 7) | Ar (2. 8. 8) |
| 4 | K (2. 8. 8. 1) | Ca (2. 8. 8. 2) | | | | | | |

This arrangement has the following features:

The elements are listed in order of increasing atomic number.

The horizontal rows are called **periods**.

The vertical columns are called **groups**. Elements that have the same number of electrons in their outermost shell fall into the same group of the Periodic Table.

The group number is the same as the number of electrons in the outermost shell (unless you label the group on the extreme right Group 0).

## Patterns in the Periodic Table

1. Elements in the same group have similar chemical properties. This is because they have the same number of electrons in their outer (valence) shell.

**Group 1** is called the **Alkali Metals**. This is because they all react with water to form alkaline solutions. They have many other similarities in their chemical reactions. ***(See Chapter 11 for further details)***

**Group 7** is called the **Halogens**. Halogen means 'salt maker'. The halogens form salts when they react with metals. For example, chlorine reacts with sodium to make sodium chloride. Sodium chloride is better known as 'common salt'. ***(See Chapter 12 for further details)***.

**Group 8/0** is called the **Noble Gases**. This group of gases used to be called the **inert** gases. Inert means chemically unreactive. These elements do not readily form compounds with other elements. Helium, although sometimes not placed in Group 8, is inert and is therefore a noble gas. The reason for the lack of reactivity is the number of electrons in the outer shell of each atom. Helium and neon atoms have full outer shells, and the rest have eight electrons in their outer shells. None of the atoms either gains or loses electrons easily.

**EXAMINER'S TIP**

Some textbooks state that the electronic configurations of the noble gases are stable and this is why they have a lack of reactivity. This is misleading, since the electronic configurations of **all** atoms are stable under normal conditions. The reasons why some atoms can readily lose electrons or gain electrons during chemical reactions are beyond the scope of International GCSE, but an Advanced Level Chemistry textbook should supply the answers.

**2.** From metal to non-metal.

From left to right across a period there is a gradual change from metal to non-metal.

For example, in Period 3, sodium, magnesium and aluminium are metals. They all conduct electricity and their oxides are basic. Phosphorus, sulfur, chlorine and argon are non-metals. They are all poor conductors of electricity and the oxides of phosphorus, sulfur and chlorine are acidic. Silicon has some properties of both a metal and a non-metal and is therefore called a **semi-metal** (or **metalloid**). Silicon is a semi-conductor of electricity and is used in computer chips for this reason. Silicon dioxide is acidic.

# **Chapter 11:** The Group 1 elements – lithium, sodium and potassium

Group 1 consists of the six metals lithium, sodium, potassium, rubidium, caesium and francium. Francium is radioactive and little is known about its chemistry other than what can be predicted from the trends within the group.

Lithium, sodium and potassium are commonly available for use in schools. They are very reactive metals and are stored under oil to prevent them from reacting with oxygen or water. These metals have the following **physical** properties:

- they are good conductors of electricity and of heat
- they are soft and can easily be cut with a knife
- they have low melting and boiling points compared with more typical metals such as iron and copper
- they have low densities (in fact all three will float on water).

Lithium, sodium and potassium have the following **chemical** properties:

- they have shiny surfaces when freshly cut with a knife, but the surface quickly becomes dull (tarnishes) as the metal reacts with oxygen in the air
- they burn in air or oxygen to form white, solid oxides. The equation for the reaction is:

$$2M + O_2 \rightarrow M_2O, \text{ where } M = Li, Na \text{ or } K$$

- they react vigorously with water to give an alkaline solution of the metal hydroxide as well as hydrogen gas. The equation for the reaction is:

$$2M + 2H_2O \rightarrow 2MOH + H_2, \text{ where } M = \text{Li, Na or K}$$

The observations made when these metals are added to water are given in the table:

| Metal | Observations |
|---|---|
| Lithium | • Moves around the surface of the water<br>• Hissing sound<br>• Bubbles of gas<br>• Gets smaller and smaller; eventually disappears |
| Sodium | • Moves around the surface of the water<br>• Hissing sound<br>• Bubbles of gas<br>• Melts into a shiny ball<br>• Gets smaller and smaller; eventually disappears |
| Potassium | • Moves around the surface of the water<br>• Hissing sound<br>• Bubbles of gas<br>• Melts into a shiny ball<br>• Burns with a lilac-coloured flame<br>• Gets smaller and smaller; eventually disappears |

- The order of reactivity of the three metals is: potassium > sodium > lithium. That is, the metals become more reactive with increasing atomic number.

Such gradual changes we call **trends**. Trends are useful to chemists as they allow predictions to be made about elements we have not observed in action, such as francium. Considering the group as a whole, the further you go down the group the more reactive the metals become. Francium is, therefore, the most reactive of the Group 1 metals.

The table shows the electronic configurations of the first three elements in Group 1.

| Element | Symbol | Atomic number | Electronic configuration |
|---|---|---|---|
| Lithium | Li | 3 | 2.1 |
| Sodium | Na | 11 | 2.8.1 |
| Potassium | K | 19 | 2.8.8.1 |

You will notice in each case that the outer shell contains only one electron. When these elements react they lose this outer electron. In order to lose an electron, the atom requires energy to overcome the electrostatic forces of attraction between the negatively charged electron and the positively charged nucleus.

Potassium is the most reactive of the three metals because less energy is required to remove the outer electron from its atom than is required for sodium or lithium.

There are two main reasons for this. As you go down the group:

- the size of the atom increases and therefore the outer electron gets further away from the nucleus, and
- the outer electron is therefore less strongly attracted to the nucleus.

# Chapter 12: The Group 7 elements – chlorine, bromine and iodine

Chlorine, bromine and iodine are the most common halogens. They are all non-metals and poisonous. Their physical states at room temperature and their colours are given in the table below:

| Halogen | Physical state at room temperature | Colour |
|---|---|---|
| Chlorine | Gas | Pale green |
| Bromine | Liquid | Red-brown (but readily evaporates to form a brown gas) |
| Iodine | Solid | Black (but sublimes when heated, to form a purple gas) |

They also react in similar ways. For example, reactions with iron are given in the table below:

| Chlorine | Bromine | Iodine |
|---|---|---|
| Hot iron wool glows brightly when chlorine passes over it. Brown smoke forms and a brown solid is left behind | Hot iron wool glows less brightly when bromine vapour is passed over it. Brown smoke and a brown solid are formed | Hot iron wool glows even less brightly when iodine vapour is passed over it. Once again, brown smoke and a brown solid are formed |

Chlorine is the most reactive of these three halogens. It reacts most vigorously with iron. Iodine is the least reactive of the three. In each case the halogen reacts with iron to form a metal halide. A halide is a **compound** of a halogen and one other element. In a **metal** halide, the other element is, obviously, a metal.

**Iron + chlorine → iron(III) chloride**

**Iron + bromine → iron(III) bromide**

**Iron + iodine → iron(III) iodide**

**Figure 12.1** *Chlorine reacting with iron wool in a gas jar.*

Reactivity **decreases** down the group because of the **increasing** size of the atoms. A halogen atom is able to attract an extra electron into its outermost shell to make eight electrons in total. It is the nucleus of each atom that attracts the extra electron.

As the atoms get bigger, their outer electron shell gets further away from the nucleus and hence the force of attraction for an electron decreases. Therefore the atom gains an extra electron less readily.

Remember! The nucleus contains protons that are positively charged. Electrons are negatively charged. Opposite charges attract.

## A further look at the reactivity of the halogens

**1.** Reactions with hydrogen are shown in the table:

| Chlorine | Bromine | Iodine |
|---|---|---|
| A mixture of hydrogen and chlorine explodes when exposed to UV radiation | A mixture of hydrogen and bromine vapour will react when heated | A mixture of hydrogen and iodine vapour will react when heated, but the reaction does not go to completion |

**EXAMINER'S TIP**

The specification does not state that you need to know the reason why the halogens become less reactive down the group, but you could be asked to suggest why this happens on the basis that you are expected to know why the Group 1 metals become more reactive down the group. This would be a typical AO2 question – application of knowledge and understanding to an unfamiliar situation.

Again, all three halogens form halides, only this time they are **hydrogen** halides.

**Hydrogen + chlorine → hydrogen chloride or $H_2 + Cl_2 \rightarrow 2HCl$**

**Hydrogen + bromine → hydrogen bromide or $H_2 + Br_2 \rightarrow 2HBr$**

**Hydrogen + iodine → hydrogen iodide or $H_2 + I_2 \rightarrow 2HI$**

**2.** Displacement reactions

Chlorine will displace bromine from an aqueous solution of a metal bromide since chlorine is more reactive than bromine. For example:

**Chlorine + potassium bromide solution → potassium chloride solution + bromine**

Chlorine will also displace iodine from an aqueous solution of a metal iodide, since chlorine is more reactive than iodine. For example:

**Chlorine + sodium iodide solution → sodium chloride solution + iodine**

Bromine will displace iodine from an aqueous solution of a metal iodide, since bromine is more reactive than iodine. For example:

**Bromine + magnesium iodide solution → magnesium bromide solution + iodine**

**In general**, a halogen will displace a **less** reactive halogen for an aqueous solution of its halide.

**EXAMINER'S TIP**

Although the specification does not require you to write ionic equations for reactions, it is much easier here to learn one equation for all the reactions that can take place. The equation is:

$X_2(aq) + 2Y^-(aq) \rightarrow 2X^-(aq) + Y_2(aq)$

If X = Cl, then Y can be Br or I

If X = Br, the Y can be I.

The halogen that is displaced dissolves in the water and colours the solution.

**Remember OILRIG:**
**O**xidation **I**s the **L**oss of electrons; **R**eduction **I**s the **G**ain of electrons.

Aqueous solutions of the halogens have the following colours:

- aqueous chlorine is very pale green, but usually appears colourless since it is often very dilute
- aqueous bromine is orange, although it will turn yellow when diluted
- aqueous iodine is brown.

## Displacement as redox reactions

In displacement reactions, the halogen molecule is gaining electrons and is therefore being reduced. The ionic half-equation for this reaction is:

$$X_2 + 2e^- \rightarrow 2X^-$$

The halide ions are losing electrons and are therefore being oxidised. The ionic half-equation for this reaction is:

$$2Y^- \rightarrow Y_2 + 2e^-$$

Since both reduction and oxidation are taking place the reaction is a **redox reaction**.

## Reactions of hydrogen chloride in solution

Hydrogen chloride, $HCl$, is a colourless gas at room temperature. It dissolves in both water and many organic solvents such as methylbenzene.

The table below lists the properties of solutions of both hydrogen chloride in methylbenzene and hydrogen chloride in water.

| Property | $HCl$ dissolved in methylbenzene | $HCl$ dissolved in water |
|---|---|---|
| Effect on blue litmus | No change | Turns it red |
| Reaction with magnesium | No change | Hydrogen evolved |
| Reaction with $Na_2CO_3$ | No change | Carbon dioxide evolved |
| Effect of electricity | Does not conduct | Hydrogen evolved at the negative electrode<br>Chlorine evolved at the positive electrode |

The following deductions in the table below can be made from the above results:

| Property | $HCl$ dissolved in methylbenzene | $HCl$ dissolved in water |
|---|---|---|
| Effect on blue litmus | Solution is not acidic | Solution is acidic |
| Reaction with magnesium | Solution is not acidic | Solution is acidic |
| Reaction with $Na_2CO_3$ | Solution is not acidic | Solution is acidic |
| Effect of electricity | Solution does not contains ions | Solution contains ions |

The hydrogen chloride molecules do not ionise when dissolved in methylbenzene, but they do when dissolved in water.

Since hydrogen is given off at the negative electrode and chlorine is given off at the positive electrode, then the ions present are hydrogen ions, $H^+$, and chloride ions, $Cl^-$ ***(see Chapter 9)***.

The equation for the ionisation is: $HCl(aq) \rightarrow H^+(aq) + Cl^-(aq)$

A solution of hydrogen chloride in water is called **hydrochloric acid**.

## Hydrochloric acid

Hydrochloric acid has the typical properties of a strongly acidic solution ***(see Chapter 21)***.

These are:

- turns litmus red
- has a low pH and hence turns Universal indicator red
- reacts with fairly reactive metals such as magnesium, zinc and iron to form a metal chloride in solution and liberate hydrogen gas

*Example:* magnesium + hydrochloric acid → magnesium chloride + hydrogen

$$Mg(s) + 2HCl(aq) \rightarrow MgCl_2(aq) + H_2$$

- reacts with metal carbonates to form a metal chloride in solution, water and carbon dioxide gas

> *Example:* calcium carbonate (marble) + hydrochloric acid → calcium chloride + water + carbon dioxide
>
> $CaCO_3(s) + 2HCl(aq) \rightarrow CaCl_2(aq) + H_2O(l) + CO_2(g)$

- reacts with bases (metal oxides and metal hydroxides) to form a metal chloride in solution and water

> *Example:* copper(II) oxide + hydrochloric acid → copper(II) chloride + water
>
> $CuO(s) + 2HCl(aq) \rightarrow CuCl_2(aq) + H_2O(l)$

# **Chapter 13:** Oxygen and oxides

## The composition of air

The table shows the approximate percentages by volume of the main gases in *unpolluted, dry* air:

| Gas | Percentage in air |
|---|---|
| nitrogen | 78 (approximately 4/5) |
| oxygen | 21 (approximately 1/5) |
| argon | 0.9 |
| carbon dioxide | 0.04 |

### Showing that air contains approximately one-fifth oxygen

#### *1. Using copper*

This apparatus can be used to find the percentage of oxygen in a sample of air:

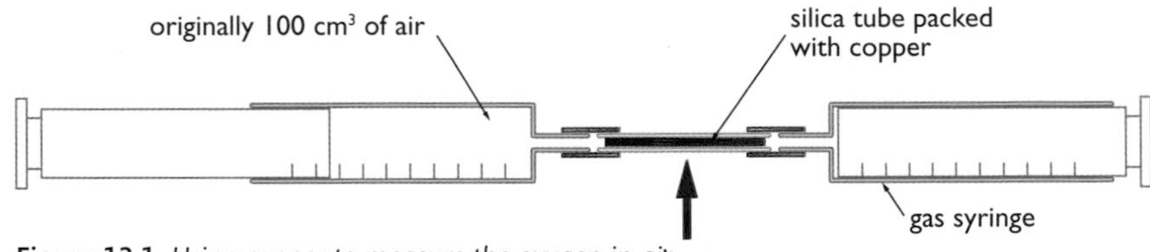


**Figure 13.1** *Using copper to measure the oxygen in air*

- set up the apparatus with 100 cm³ of air in one of the gas syringes
- heat the copper at one end of the silica tube using a blue Bunsen flame
- pass the air backwards and forwards over the copper
- as the volume of gas in the syringes decreases, move the Bunsen flame along the tube so that it is always heating fresh copper
- stop heating when the volume of gas in the syringe stops decreasing. The copper is reacting with the oxygen in the air to form black copper(II) oxide:

$$2Cu(s) + O_2(g) \rightarrow 2CuO(s)$$

The final volume of air in the syringe will be approximately 79 cm³, showing that 21 cm³ has reacted. That is, 21% has been used up. Therefore the air was 21% oxygen.

> **EXAMINER'S TIP**
> It is important that enough copper is available to react with *all* the oxygen in the sample of air. This can only be shown by some of the copper remaining unchanged at the end of the experiment.

## 2. Using iron

Place wet iron filings in the end of a burette and set up the apparatus as shown in the diagram. Over several days the water will rise up the burette and reach a constant level. This is because the iron reacts with the oxygen in the air. Take the initial and final readings of the water level in the burette.

The volume of air at the start = (50 – initial burette reading)

Volume of oxygen used (reacted with iron) = (Initial reading – final reading)

The percentage of oxygen in air can be calculated using the equation:

$$\text{percentage of oxygen} = \frac{\text{volume of oxygen used}}{\text{volume of air at start}} \times 100$$

**Figure 13.2** *Using iron to measure the oxygen in air.*

## 3. Using phosphorus

A similar experiment can be done with a piece of white phosphorus.

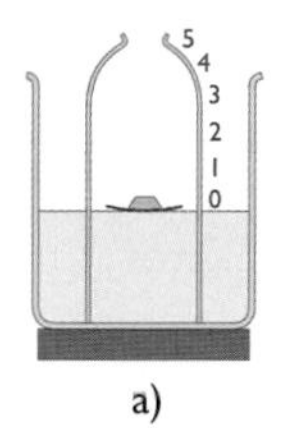

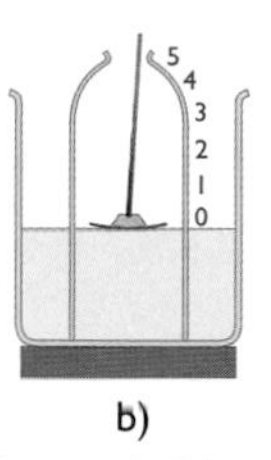

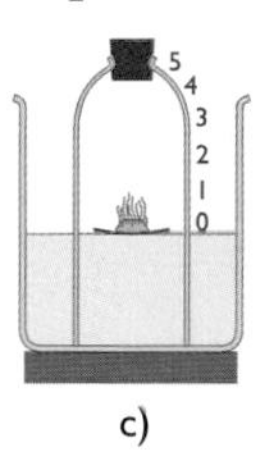

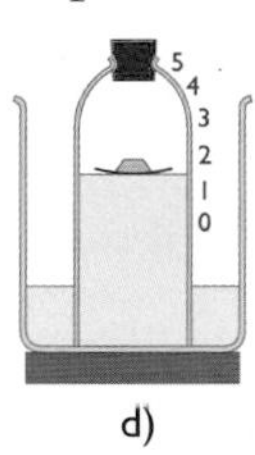

a) b) c) d)

**Figure 13.3** ***a)*** *Levels of water inside and outside the tube equal.* ***b)*** *Levels equal; phosphorus touched by hot metal rod.* ***c)*** *Levels equal; phosphorus starts burning.* ***d)*** *Levels inside higher: phosphorus stops burning.*

The equation for the reaction is:

$$4P(s) + 5O_2(g) \rightarrow 2P_2O_5(s)$$

## The laboratory preparation of oxygen

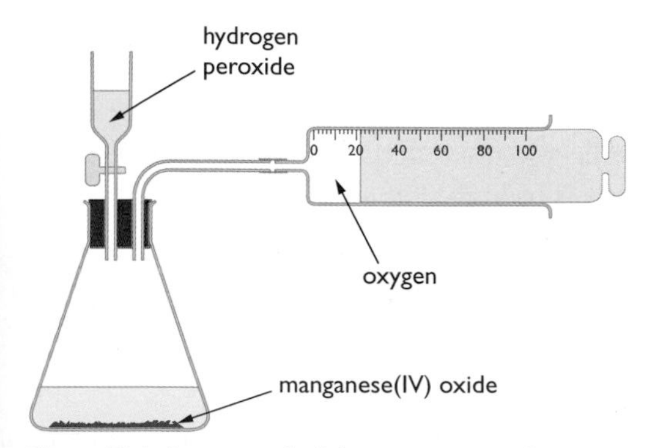


**Figure 13.4** *Apparatus for laboratory preparation of oxygen.*

Hydrogen peroxide, $H_2O_2$, decomposes slowly to form water and oxygen. The speed of the decomposition is increased by adding solid manganese dioxide, $MnO_2$, which acts as a catalyst for the reaction. The oxygen can also be collected over water. Since oxygen is not very soluble in water, very little is lost. An aqueous solution of hydrogen peroxide is used in the laboratory preparation:

$$2H_2O_2(aq) \rightarrow 2H_2O(l) + O_2(g)$$

## Reactions of oxygen

Oxygen is a reactive gas. Any substance that burns in air will burn more brightly and more vigorously in oxygen. The table below contains descriptions of the reactions of three elements when heated in oxygen:

| Element | Observations | Equation for reaction |
|---|---|---|
| magnesium | burns with bright, white flame to form a white powder | $2Mg(s) + O_2(g) \rightarrow 2MgO(s)$ |
| carbon | burns with a yellow-orange flame to form a colourless gas | $C(s) + O_2(g) \rightarrow CO_2(g)$ |
| sulfur | burns with a blue flame to form a colourless gas | $S(s) + O_2(g) \rightarrow SO_2(g)$ |

Magnesium oxide, MgO, is a basic oxide ***(see Chapter 21)***. It is very slightly soluble in water and a saturated solution will have a pH of about 10. It reacts with water to form a solution of magnesium hydroxide, $Mg(OH)_2$:

$$MgO(s) + H_2O(l) \rightarrow Mg(OH)_2(aq)$$

Both carbon dioxide, $CO_2$, and sulfur dioxide, $SO_2$, are acidic oxides. They dissolve in water to form acidic solutions:

$$CO_2(g) + H_2O(l) \rightarrow H_2CO_3(aq)$$
**carbonic acid (pH approximately 5–6)**

$$SO_2(g) + H_2O(l) \rightarrow H_2SO_3(aq)$$
**sulfurous acid (pH approximately 3–4)**

Carbon dioxide is only slightly soluble in water, but sulfur dioxide is very soluble.

To summarise:

- Oxides of metals are basic. If they dissolve in water (and most do not) they form alkaline solutions.
- Oxides of non-metals are often acidic. If they dissolve in water they form acidic solutions. Some oxides of non-metals are neutral, e.g. carbon monoxide.

## The laboratory preparation of carbon dioxide

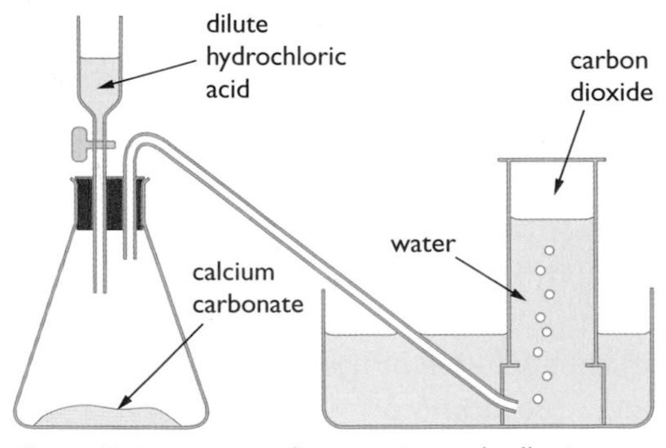


**Figure 13.5** *Apparatus for preparing and collecting carbon dioxide.*

The reaction between any metal carbonate and an acid will produce carbon dioxide. Calcium carbonate is the most commonly used metal carbonate in the laboratory preparation of carbon dioxide. The most convenient form of calcium carbonate to use is marble chips. They are very easy to handle and the reaction is not too fast, so the carbon dioxide is produced at a rate that makes it easy to collect.

$$CaCO_3(s) + 2HCl(aq) \rightarrow CaCl_2(aq) + H_2O(l) + CO_2(g)$$

Carbon dioxide is not very soluble in water so it can be collected over water without much being lost. It can also be collected by **downward** delivery in air, as it is more dense than air.

Carbon dioxide is also produced when most metal carbonates are heated. The table opposite describes the **thermal decomposition** (i.e. breakdown by heating) of some metal carbonates.

| Metal carbonate | Observations | Equation for reaction |
|---|---|---|
| copper(II) carbonate | green to black, see Figure 13.6 | $CuCO_3(s) \rightarrow CuO(s) + CO_2(s)$ |
| magnesium carbonate | no observable change – stays white | $MgCO_3(s) \rightarrow MgO(s) + CO_2(g)$ |
| calcium carbonate | no observable change – stays white | $CaCO_3(s) \rightarrow CaO(s) + CO_2(g)$ |
| zinc carbonate | white to yellow when hot and white again when cold | $ZnCO_3(s) \rightarrow ZnO(s) + CO_2(g)$ |
| sodium carbonate | no observable change – stays white | does not decompose in a Bunsen flame |

## Uses of carbon dioxide

- Making carbonated drinks. Although carbon dioxide is not very soluble in water at normal atmospheric pressure, it becomes much more soluble when put under pressure. Use of this is made in making 'fizzy' drinks such as tonic water, soda water, coca-cola, and so on.
- In fire extinguishers. Carbon dioxide does not support combustion and is more dense than air, so it 'sits' on top of the burning fuel and prevents oxygen from getting to it.

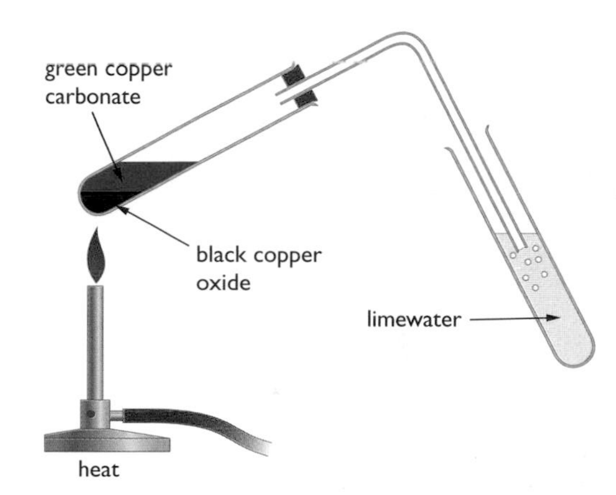


**Figure 13.6** *Test for $CO_2$: limewater turns milky in the presence of carbon dioxide.*

## Carbon dioxide and climate change

Carbon dioxide is a greenhouse gas since it absorbs infra-red radiation given off from the Earth's surface. In recent years the amount of carbon dioxide in the Earth's atmosphere has increased. One reason for this is that we are burning more fossil fuels and fuels made from fossil fuels (such as petrol) than ever before. Some scientists think that this increase in carbon dioxide in the atmosphere is contributing to climate change. Other scientists believe that recent changes in climatic conditions are just part of natural changes that have occurred since the Earth was formed. The debate is continuing.

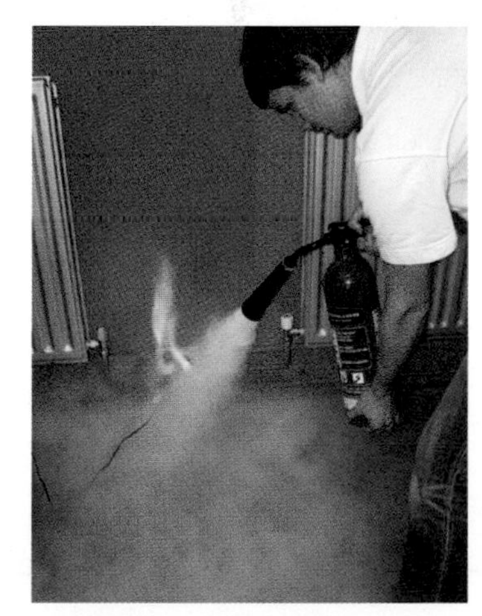

**Figure 13.7** *Carbon dioxide extinguishers can be used to put out electrical fires or fires involving combustible solids and liquids.*

## Acid rain

Rain is naturally acidic because of dissolved carbon dioxide. The pH of 'normal' rain is about 6. However, certain pollutant gases such sulfur dioxide and oxides of nitrogen can cause the pH to drop significantly below this value. Rain with a pH lower than 6 is called 'acid rain'. Acid rain causes a number of problems such as:

- leaching vital minerals out of the soil and hence causing trees to die
- lowering the pH of lakes and rivers so that fish and other aquatic life cannot survive
- weathering of buildings and structures made out of limestone, marble (both forms of calcium carbonate) and iron.

# Chapter 14: Hydrogen and water

## Reaction of metals with dilute acids to produce hydrogen

Metals above hydrogen in the reactivity series ***(see Chapter 15)*** react with both dilute hydrochloric acid, $HCl(aq)$, and dilute sulfuric acid, $H_2SO_4(aq)$, to form a salt and hydrogen:

**metal + dilute hydrochloric acid → metal chloride + hydrogen**

**metal + dilute sulfuric acid → metal sulfate + hydrogen**

The table describes the reactions of some metals with each of these acids:

| Metal | Observation with dil. $HCl$ or dil. $H_2SO_4$ | Equations |
|---|---|---|
| magnesium | • bubbles of gas<br>• magnesium disappears<br>• reaction mixture gets hot<br>• colourless solution formed | $Mg(s) + 2HCl(aq) \rightarrow MgCl_2(aq) + H_2(g)$<br>$Mg(s) + H_2SO_4(aq) \rightarrow MgSO_4(aq) + H_2(g)$ |
| aluminium | • slow to start reacting when cold, but bubbles form when heated<br>• aluminium disappears<br>• colourless solution formed | $2Al(s) + 6HCl(aq) \rightarrow 2AlCl_3(aq) + 3H_2(g)$<br>$2Al(s) + 3H_2SO_4(aq) \rightarrow Al_2(SO_4)_3(aq) + 3H_2(g)$ |
| zinc | • bubbles of gas<br>• zinc disappears<br>• colourless solution formed | $Zn(s) + 2HCl(aq) \rightarrow ZnCl_2(aq) + H_2(g)$<br>$Zn(s) + H_2SO_4(aq) \rightarrow ZnSO_4(aq) + H_2(g)$ |
| iron | • bubbles of gas<br>• iron disappears<br>• pale green solution formed | $Fe(s) + 2HCl(aq) \rightarrow FeCl_2(aq) + H_2(g)$<br>$Fe(s) + H_2SO_4(aq) \rightarrow FeSO_4(aq) + H_2(g)$ |

## The combustion of hydrogen

Hydrogen burns when heated in air or oxygen to form water:

$$2H_2(g) + O_2(g) \rightarrow 2H_2O(l)$$

The product will be formed initially as water vapour, but if cooled it can be condensed to form water.

**Test for hydrogen**: A mixture of hydrogen and air/oxygen will explode when ignited by a spark of a flame and water is the product (as in the equation above). This provides the characteristic test for hydrogen. The test is as follows: Mix a test tube full of hydrogen with air and place the flame of a lighted spill at the mouth of the test tube. A squeaky pop will be heard.

**Test for water**: The product from the burning of hydrogen in oxygen can be shown to be pure water through a combination of two tests.

This test does not confirm that the liquid is **pure** water, only that water is present.

1. Add the liquid (from condensing the water vapour) to anhydrous copper(II) sulfate. The white powder will turn blue and hydrated copper(II) sulfate is formed:

$$\underset{\text{white}}{CuSO_4(s)} + 5H_2O(l) \rightarrow \underset{\text{blue}}{CuSO_4.5H_2O(s)}$$

**Figure 14.1** *Adding water to anhydrous copper(II) sulfate.*

2. Boil the liquid and measure its boiling point. It will be 100°C. Alternatively, you can measure its freezing point. It will be 0°C.

These two tests together confirm that the liquid is pure water. If the water contains impurities, its boiling point will be above 100°C and its freezing point will be less than 0°C.

# **Chapter 15:** Reactivity series

The table below summarises the reactions of some metals with water and acids. The metals are placed in order of decreasing reactivity, i.e. most reactive at the top.

<table>
<tr><th>Metal</th><th>Reaction with cold water or steam</th><th>Reaction with dilute acids</th></tr>
<tr><td>Potassium</td><td rowspan="4">React vigorously with cold water to form a solution of the metal hydroxide and hydrogen</td><td rowspan="3">React very vigorously to form metal salt and hydrogen</td></tr>
<tr><td>Sodium</td></tr>
<tr><td>Lithium</td></tr>
<tr><td>Calcium</td><td rowspan="7">React with decreasing vigour as the series is descended, to form the metal salt and hydrogen</td></tr>
<tr><td>Magnesium</td><td>Reacts very slowly with cold water but burns in steam to form magnesium oxide and hydrogen</td></tr>
<tr><td>Aluminium</td><td rowspan="3">Do not react with cold water but do react with steam, without burning, to form metal oxide and hydrogen</td></tr>
<tr><td>Zinc</td></tr>
<tr><td>Iron</td></tr>
<tr><td>Tin</td><td rowspan="6">Do not react with either cold water or steam</td></tr>
<tr><td>Lead</td></tr>
<tr><td>Copper</td><td rowspan="4">Do not react</td></tr>
<tr><td>Silver</td></tr>
<tr><td>Gold</td></tr>
<tr><td>Platinum</td></tr>
</table>

You may have read that aluminium is resistant to corrosion and therefore you may be surprised to see aluminium placed as high as it is in the reactivity series. In fact, aluminium is a reactive metal; as soon as freshly prepared aluminium is exposed to air it reacts with oxygen to form a coating of aluminium oxide. This surface coating of aluminium oxide is unreactive and prevents the metal from showing its true reactivity.

## Using the reactivity series

### Reducing metal oxides

A metal oxide is a compound of a metal and oxygen. When the oxygen is removed from a metal oxide, the oxide is said to be **reduced**. The substance that carries out the reduction is called the **reducing agent**.

A metal oxide can be reduced by heating it with a metal (the reducing agent) that is higher in the reactivity series. For example, it is possible to reduce copper(II) oxide by heating it with magnesium:

$$CuO(s) + Mg(s) \rightarrow Cu(s) + MgO(s)$$

The reducing agent in this reaction is magnesium.

The reaction is best carried out in a crucible made of heat-resistant porcelain. Both the metal and the oxide should be in powdered form. The reaction is highly **exothermic** (i.e. it produces a lot of heat) and therefore the mixture will glow very brightly. The final mixture will contain magnesium oxide (a white powder) and copper (a red-brown powder).

The table below shows the reactions, or lack of reactions, when some metals and metal oxides are heated together:

| Mixture | Products | Equation |
|---|---|---|
| iron(III) oxide and aluminium | iron and aluminium oxide | $Fe_2O_3(s) + 2Al(s) \rightarrow 2Fe(s) + Al_2O_3(s)$ |
| sodium oxide and magnesium | no reaction | – |
| silver oxide and copper | silver and copper(II) oxide | $Ag_2O(s) + Cu(s) \rightarrow 2Ag(s) + CuO(s)$ |
| zinc oxide and calcium | zinc and calcium oxide | $ZnO(s) + Ca(s) \rightarrow Zn(s) + CaO(s)$ |
| lead(II) oxide and silver | no reaction | – |

## The position of carbon in the reactivity series

Carbon can also combine with oxygen (to form carbon dioxide) and therefore can be placed in the reactivity series. Carbon is placed between aluminium and zinc, because it can reduce zinc oxide (and, therefore, the oxides of all the other metals below zinc) but it cannot reduce aluminium oxide.

When carbon reduces a metal oxide to a metal, the carbon is converted into carbon dioxide. For example:

$$2Fe_2O_3(s) + 3C(s) \rightarrow 4Fe(s) + 3CO_2(g)$$

## Displacement of metals from their salts

Any metal will displace another metal that is below it in the reactivity series from a solution of one of its salts. For example, zinc displaces copper from copper(II) sulfate solution:

$$Zn(s) + CuSO_4(aq) \rightarrow ZnSO_4(aq) + Cu(s)$$

Salts of metals include chlorides, nitrates and sulfates.

It is important to know which salts are soluble in water. The following is a solubility list for the common metals:

**Chlorides** All common metal chlorides are soluble in water except silver chloride and lead(II) chloride

**Nitrates** All metal nitrates are soluble in water

**Sulfates** All common metal sulfates are soluble in water except barium sulfate, calcium sulfate and lead(II) sulfate

The table below describes the reactions, or lack of reactions, when some metals are added to solutions of metal salts:

| Mixture | Products | Equation |
|---|---|---|
| magnesium and iron(II) sulfate | Magnesium sulfate and iron | $Mg(s) + FeSO_4(aq) \rightarrow MgSO_4(aq) + Fe(s)$ |
| zinc and sodium chloride | No reaction | – |
| lead and silver nitrate | Lead(II) nitrate and silver | $Pb(s) + 2AgNO_3(aq) \rightarrow Pb(NO_3)_2(aq) + 2Ag(s)$ |
| copper and calcium chloride | No reaction | – |
| iron and copper(II) sulfate | Iron(II) sulfate and copper | $Fe(s) + CuSO_4(aq) \rightarrow FeSO_4(aq) + Cu(s)$ |

## The position of hydrogen in the reactivity series

Hydrogen, although not a metal, is included in the reactivity series because it, like metals, can be displaced from aqueous solution, only this time the solution is an acid.

Hydrogen is placed between lead and copper. All metals above hydrogen in the reactivity series can displace hydrogen gas from dilute hydrochloric acid and dilute sulfuric acid. Those metals below hydrogen in the reactivity series cannot displace hydrogen from solutions of acids.

# Rusting of iron

Rusting is a chemical reaction between iron, water and oxygen. Water and oxygen must both be present for rusting to occur. Rusting takes place faster if there are also electrolytes such as sodium chloride in the water. This is why rusting takes place much faster in sea water.

Rust is a complicated compound, but can be represented by the formula $Fe_2O_3.xH_2O$, where $x$ is a variable number. The iron atoms have been oxidised to iron(III) ions ($Fe^{3+}$) by the loss of electrons.

## Preventing rusting

The rusting of iron and steel can be prevented in a number of ways. The obvious way is to prevent the iron from coming into contact with water and oxygen. This can be done in a number of ways by coating the iron with:

- grease
- oil
- paint
- plastic
- a metal less reactive than iron, such as tin.

**EXAMINER'S TIP**

Many metals corrode when exposed to air, but it is only the corrosion of iron that is referred to as rusting.

However, if the coatings are washed away or scratched, the iron is once again exposed to water and oxygen and will rust.

### *Sacrificial protection of iron*

Iron can be prevented from rusting by using what we know about the reactivity series. Zinc is above iron in the reactivity series, so zinc reacts more readily than iron.

Galvanised iron is iron that is coated with a layer of zinc. To begin with, the coating will protect the iron. If the coating is damaged or scratched, the iron is still protected from rusting. This is because zinc is more reactive than iron and so it reacts and corrodes instead of the iron.

If zinc blocks are attached to the hulls of ships, they will corrode instead of the hull. The zinc is called a sacrificial anode.

# Chapter 16: Tests for ions and gases

## Identification of cations (positive ions)

### Identifying metal cations by using flame tests

The principle here is that the salts of some metals will impart a colour to a non-luminous Bunsen flame.

**Test**

1. The technique is first of all to clean the end of a piece of platinum or nichrome wire by dipping it into clean hydrochloric acid and then placing it in a roaring Bunsen flame. This procedure should be repeated until the wire no longer produces a colour in the flame.
2. The end of the wire should then be dipped into fresh hydrochloric acid and then into the solid sample under test.
3. The end of the wire should then be placed into a non-roaring, non-luminous Bunsen flame.

**Result:** In the table below are some of the common metal cations that can be tested in this way:

| Metal cation | Colour of flame |
|---|---|
| Lithium, $Li^+$ | Red |
| Potassium, $K^+$ | Lilac |
| Calcium, $Ca^{2+}$ | Brick red |
| Sodium, $Na^+$ | Yellow / orange |

### Identifying the ammonium ion, $NH_4^+$

**Test:** Add aqueous sodium hydroxide to the solid, or solution, under test and warm the mixture.

**Result:** If ammonium ions are present then a pungent-smelling gas is produced. The gas produced turns damp red litmus paper blue. It is ammonia, $NH_3$.

**Equation:**

| $NH_4^+(aq)$ | + | $OH^-$ | $\rightarrow$ | $NH_3$ | + | $H_2O$ |
|---|---|---|---|---|---|---|
| ammonium ions (from the solution being tested) | | hydroxide ions (from the sodium hydroxide added) | | | | |

### Identifying metal cations by using precipitation reactions

Most metal hydroxides are insoluble and hence can be precipitated from aqueous solutions of metal salts by adding an aqueous solution of sodium hydroxide.

The technique is to add the reagent (i.e. the aqueous sodium hydroxide) a drop at a time to form the precipitate.

**EXAMINER'S TIP**

The ionic equations given here apply to any combinations of metal salt and sodium hydroxide. Although there is no requirement in the specification to learn ionic equations for these reactions, it is far easier to do so than to try to remember or work out individual equations for each separate combination of metal salt and sodium hydroxide.

Some common metal cations that can be identified using these reagents are shown in the table below:

| Metal cation | Observation(s) with aq. sodium hydroxide | Equation for reaction |
|---|---|---|
| copper(II), $Cu^{2+}$ | blue precipitate | $Cu^{2+}(aq) + 2OH^-(aq) \rightarrow Cu(OH)_2(s)$ |
| iron(II), $Fe^{2+}$ | green precipitate | $Fe^{2+}(aq) + 2OH^-(aq) \rightarrow Fe(OH)_2(s)$ |
| iron(III), $Fe^{3+}$ | brown precipitate | $Fe^{3+}(aq) + 3OH^-(aq) \rightarrow Fe(OH)_3(s)$ |

# Identification of anions (negative ions)

## Halide ions by precipitation with silver nitrate solution

**Test:** To an aqueous solution of the solid under test, add some dilute nitric acid followed by a few drops of silver nitrate solution.

**Results:**

| Halide ion present | Observation | Equation for reaction |
|---|---|---|
| chloride ion, $Cl^-$ | white precipitate (of silver chloride, AgCl) | $Ag^+(aq) + Cl^-(aq) \rightarrow AgCl(s)$ |
| bromide ion, $Br^-$ | cream preclpltate (of silver bromide, AgBr) | $Ag^+(aq) + Br^-(aq) \rightarrow AgBr(s)$ |
| iodide ion, $I^-$ | yellow precipitate (of silver iodide, AgI) | $Ag^+(aq) + I^-(aq) \rightarrow AgI(s)$ |

## Sulfate ions, $SO_4^{2-}$, by precipitation with barium chloride solution

**Test:** To an aqueous solution of the solid under test, add dilute hydrochloric acid followed by a few drops of barium chloride solution.

**Result:** White precipitate (of barium sulfate)

**Equation:** $Ba^{2+}(aq) + SO_4^{2-}(aq) \rightarrow BaSO_4(s)$

## Carbonate ions using dilute acid

**Test:** To either the solid, or an aqueous solution of the solid, under test add dilute hydrochloric (or nitric) acid.

**Result:** Bubbles of gas. The gas produced turns limewater milky.

**Equation:** $CO_3^{2-}(aq) + 2H^+(aq) \rightarrow CO_2(g) + H_2O(l)$

## Tests for gases

| Name of gas | Test | Result if positive | Equation |
|---|---|---|---|
| hydrogen | mix with air and ignite | burns with a 'squeaky pop' | $2H_2(g) + O_2(g) \rightarrow 2H_2O(l)$ |
| oxygen | insert glowing spill 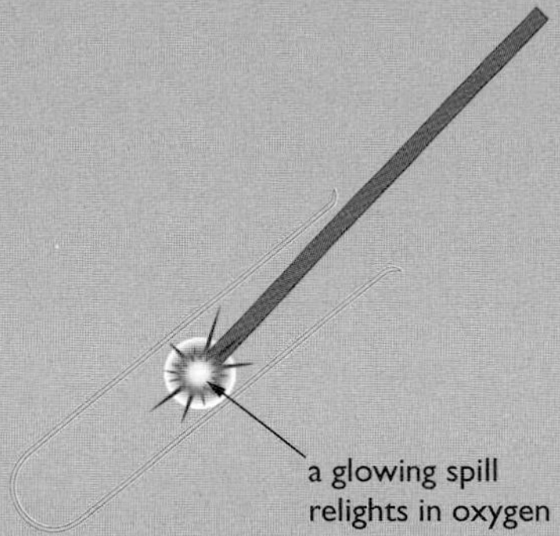 | spill relights | |
| carbon dioxide | bubble through limewater 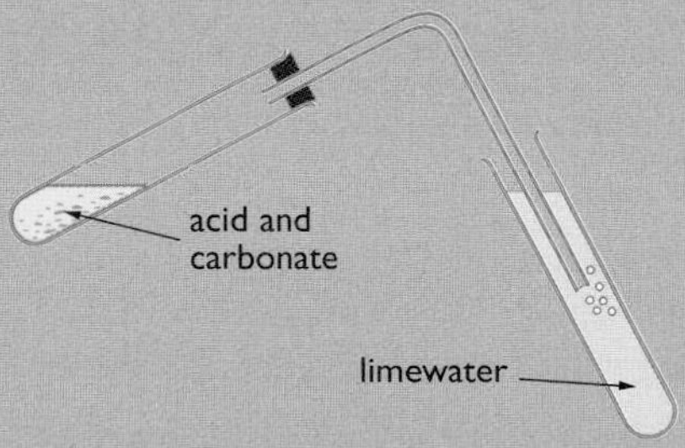 | limewater turns milky | $Ca(OH)_2(aq) + CO_2(g) \rightarrow CaCO_3(s) + H_2O(l)$<br>(the white solid, $CaCO_3$, turns the limewater milky) |
| chlorine | damp litmus paper<br>OR<br>moist starch-iodide paper | litmus paper turns white<br><br>turns blue | $Cl_2(g) + 2I^-(aq) \rightarrow 2Cl^-(aq) + I_2(aq)$<br>(the iodine formed turns the starch blue) |
| ammonia | damp red litmus paper<br>OR<br>damp universal indicator (pH) paper 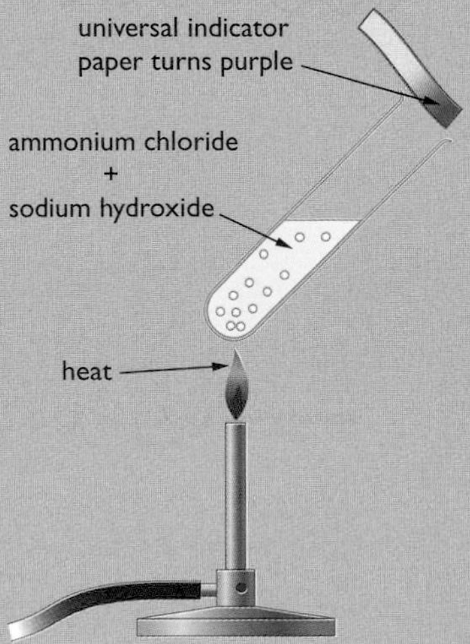 | turns blue<br><br>turns purple | $NH_3(g) + H_2O(l) \rightleftharpoons NH_4^+(aq) + OH^-(aq)$<br>(the hydroxide ions, $OH^-$, turn the litmus blue) |

## Examination Questions

**1** Use the Periodic Table on p. 119 to help answer this question.

*a)* Which number increases from 3 to 10 in Period 2? *(1)*

*b)* Which number increases from 11 to 204 in Group 3? *(1)*

*c)* Which group contains elements whose ions all have a 1+ charge? *(1)*

*d)* Which group contains elements whose ions have a 2- charge? *(1)*

*e)* Give the number of a period that contains transition metals. *(1)*

***(Total 5 marks)***

**2** A mixture contains an insoluble compound and a soluble compound.

The mixture is separated by adding hot water and then filtering.

This produces a **white** solid, **A**, and a **green** solution, **B**.

The white solid and the green solution were tested to find out what they were. The tables show the tests used and the results.

| Tests on white solid A | |
|---|---|
| **Test** | **Result** |
| Carry out flame test | The flame was coloured brick red |
| Add dilute hydrochloric acid<br>Test the gas produced | Bubbles seen<br>Found to be carbon dioxide |

*a)* (i) **Identify** the cation in solid **A**. *(1)*

(ii) The gas produced is carbon dioxide. State the test for carbon dioxide and the result of this test. *(2)*

(iii) **Identify** the anion in solid **A**. *(1)*

| Tests on green solution B | |
|---|---|
| **Test** | **Result** |
| Add sodium hydroxide solution | Green precipitate |
| Add dilute nitric acid<br>Then add silver nitrate solution | No change<br>No change |
| Add barium chloride solution<br>Then add dilute hydrochloric acid | White precipitate<br>No change |

*b)* (i) State the **formula** of the cation in solution **B**. *(1)*

(ii) State the **name** of the green precipitate. *(1)*

(iii) **Identify** the anion in solution **B**. *(1)*

(iv) State the **formula** of the white precipitate. *(1)*

*c)* There are three anions that give a precipitate when dilute nitric acid and silver nitrate solution are added. Name **two** of these anions. *(2)*

*d)* (i) State the **formula** of solid **A**. *(1)*

(ii) State the **formula** of the compound in solution **B**. *(1)*

***(Total 12 marks)***

**3** The pie chart shows the approximate percentages by volume of gases in dry air.

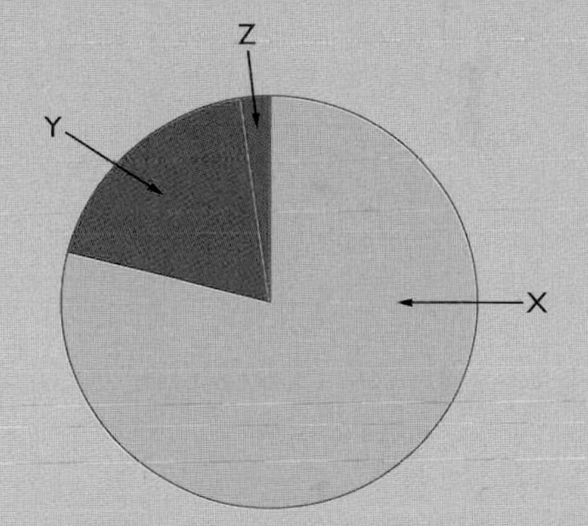


*a)* Which part of the pie chart (**X**, **Y** or **Z**) represents nitrogen? *(1)*

*b)* What is the approximate percentage of oxygen in dry air? *(1)*

*c)* What is the test for oxygen gas? *(2)*

*d)* Name the gas in dry air that is formed by the complete combustion of methane, $CH_4$ *(1)*

***(Total 5 marks)***

**4** The table gives the colours of some indicators at different pH values.

<table>
<tr><th rowspan="2">Indicator</th><th colspan="7">pH</th></tr>
<tr><th>1</th><th>3</th><th>5</th><th>7</th><th>9</th><th>11</th><th>13</th></tr>
<tr><td>litmus</td><td colspan="3">◄ red ►</td><td>purple</td><td colspan="3">◄ blue ►</td></tr>
<tr><td>phenolphthalein</td><td colspan="5">◄ colourless ►</td><td colspan="2">◄ pink ►</td></tr>
<tr><td>methyl orange</td><td colspan="2">◄ red ►</td><td colspan="5">◄ yellow ►</td></tr>
</table>

*a)* (i) Use the table to find the pH of a solution in which litmus is red **and** methyl orange is yellow. *(1)*

(ii) Litmus is purple in sodium chloride solution. What colour is phenolphthalein in sodium chloride solution? *(1)*

**b)** A student was investigating the neutralisation of aqueous ammonia using hydrochloric acid.

She placed 25 $cm^3$ of aqueous ammonia in a conical flask and added a few drops of litmus.

She then slowly added hydrochloric acid to the mixture in the flask.

The indicator turned purple after she had added 15 $cm^3$ of hydrochloric acid.

The word equation for the reaction is: ammonia + hydrochloric acid → ammonium chloride

(i) Write a chemical equation for the reaction of ammonia with hydrochloric acid. *(2)*

(ii) Describe a chemical test to show that the solution obtained contains ammonium ions. Give the result of the test. *(3)*

(iii) The student used the same original solutions of aqueous ammonia and hydrochloric acid to make a pure sample of ammonium chloride crystals. Describe how she could do this. *(3)*

**c)** (i) Lead(II) chloride is insoluble. Name two solutions that react together to make lead(II) chloride. *(2)*

(ii) Write a **word** equation for this reaction. *(1)*

***(Total 13 marks)***

**5** This question is about the reactions of the metals calcium, iron and zinc.

**a)** Samples of each of the powdered metals were placed in separate beakers of water. Only calcium reacted immediately.

Describe **two** observations that could be made during the reaction of calcium with water. Write a chemical equation for the reaction. *(3)*

**b)** A reaction occurred when powdered zinc was heated in steam.

Name the zinc compound formed. Write a chemical equation for the reaction. *(2)*

**c)** Some powdered zinc was added to a solution of iron(II) sulfate.

(i) Write an ionic equation to show the reaction that occurs. *(1)*

(ii) State the type of reaction occurring. *(1)*

**d)** Iron rusts slowly in the presence of water. Name one other substance that must be present for iron to rust. *(1)*

**e)** Galvanising is one method used to prevent iron from rusting.

(i) Describe how a sheet of iron is galvanised. *(1)*

(ii) A sheet of galvanised iron was scratched and left in the rain. The exposed iron did not rust. Explain why. *(2)*

***(Total 11 marks)***

**6** The reactivity of metals can be compared by their reactions with dilute hydrochloric acid.

Three different metals are added to separate test tubes containing this acid.

The diagram shows bubbles of hydrogen gas forming when a piece of zinc is added to dilute hydrochloric acid.

**a)** Complete the diagram to show the bubbles forming in the other two test tubes. *(2)*

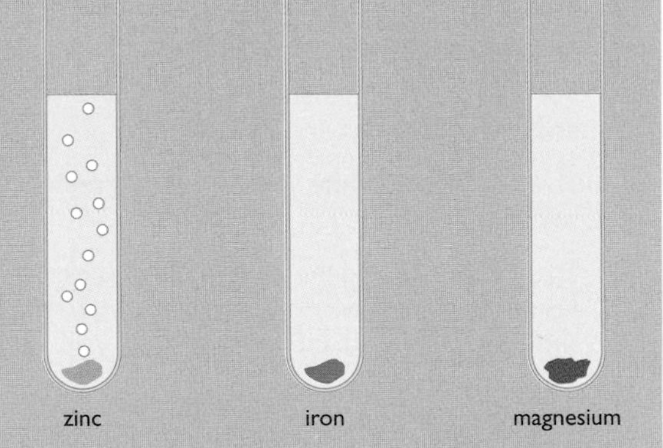


**b)** Write a word equation for the reaction between zinc and dilute hydrochloric acid. *(1)*

**c)** Name **one** metal that does not form bubbles when it is added to dilute hydrochloric acid. *(1)*

**d)** Identify **two** substances, other than acids, that can be used in reactions to compare the reactivity of metals. *(2)*

***(Total 6 marks)***

**7** A student tests a solution to see if it contains $CO_3^{2-}$ ions. The first part of this test involves this reaction:

$2H^+(\_\_) + CO_3^{2-}(aq) \rightarrow H_2O(\_\_) + CO_2(\_\_)$

**a)** One state symbol is given in the equation. Write the other state symbols in the spaces provided. *(3)*

**b)** Name a reagent that can be used to provide the $H^+$ ions in the reaction. *(1)*

**c)** State the name for each of the following formulae: $CO_3^{2-}$ and $CO_2$. *(2)*

**d)** The second part of the test involves using $Ca(OH)_2$ to detect the $CO_2$.

(i) What is the chemical name for $Ca(OH)_2$? *(1)*

(ii) The $Ca(OH)_2$ is dissolved in water to make a solution when doing the test for $CO_2$. What is the common name for this solution? *(1)*

(iii) What is **seen** during this test for $CO_2$? *(1)*

(iv) Complete the chemical equation for the reaction between these two substances.

$Ca(OH)_2 + CO_2 \rightarrow$ ______ + ______ *(2)*

*e)* $CO_2$ is present in air. What effect does it have on rain water? *(1)*

*(Total 12 marks)*

**8** Sodium is a very reactive metal. It floats on water and reacts rapidly with water.

A small piece of sodium is placed in a trough of water. A reaction takes place and hydrogen gas is given off.

*a)* (i) Give **two** observations, other than the sodium floating, that you could make during the reaction. *(2)*

(ii) Write a word equation for the reaction. *(1)*

(iii) Universal indicator is added to the water in the trough. State what colour it turns and explain why. *(2)*

*b)* A piece of platinum wire is dipped into the solution in the trough and then held in a roaring Bunsen flame. The Bunsen flame becomes coloured.

(i) What colour does the flame become? *(1)*

(ii) What name is given to this method of identification? *(1)*

*c)* A piece of sodium is heated in a Bunsen flame. The sodium catches fire and reacts with the oxygen in the air. The product is sodium oxide.

(i) The diagrams show the electron arrangement in an atom of sodium and an atom of oxygen.

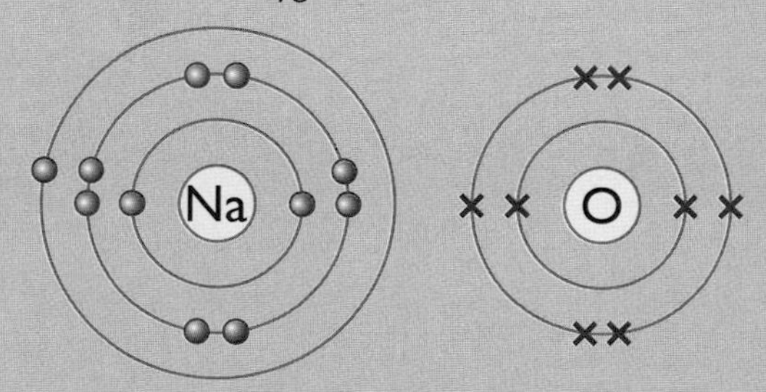


Sodium oxide contains ionic bonds. Describe what happens, in terms of electrons, when sodium reacts with oxygen. *(3)*

(ii) Draw circles round the symbols that represent the two ions produced. *(2)*

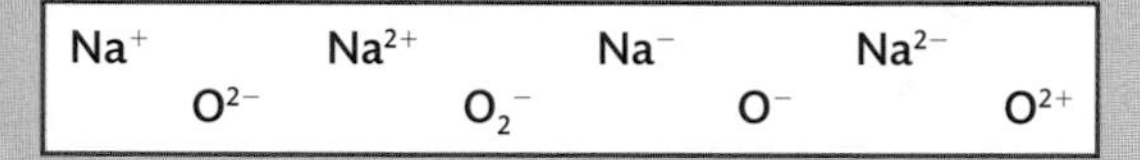


*(Total 12 marks)*

**9** Three of the elements in Group 7 of the Periodic Table are chlorine, bromine and iodine.

*a)* Give the electronic configuration of chlorine. *(1)*

*b)* How many electrons are there in the outer shell of an atom of iodine? *(1)*

*c)* Bromine reacts with hydrogen to form hydrogen bromide. The chemical equation for the reaction is: $Br_2(g) + H_2(g) \rightarrow 2HBr(g)$

Describe the colour change occurring during the reaction. *(2)*

*d)* Hydrogen bromide and hydrogen chloride have similar chemical properties.

(i) A sample of hydrogen bromide is dissolved in water. A piece of blue litmus paper is placed in the solution. State, with a reason, the final colour of the litmus paper. *(2)*

(ii) A sample of hydrogen bromide is dissolved in methylbenzene. A piece of blue litmus paper Is placed in the solution. State, with a reason, the final colour of the litmus paper. *(2)*

*(Total 8 marks)*

**10** Calcium and magnesium are metals in Group 2 of the Periodic Table.

*a)* (i) How many electrons are there in the outer shell of an atom of calcium? *(1)*

(ii) Write the electronic configuration of an atom of magnesium. *(1)*

*b)* A student adds a piece of calcium to some cold water in a beaker. The products of the reaction are calcium hydroxide and hydrogen. Some of the calcium hydroxide dissolves in the water and some does not.

(i) Describe **two** observations that the student could make during the reaction. *(2)*

(ii) Give the formula of calcium hydroxide. *(1)*

(iii) When the reaction is complete, a piece of litmus paper is added to the solution in the beaker. State the final colour of the litmus paper and what this colour indicates about the solution. *(2)*

c) The diagram shows apparatus for reacting magnesium with steam.

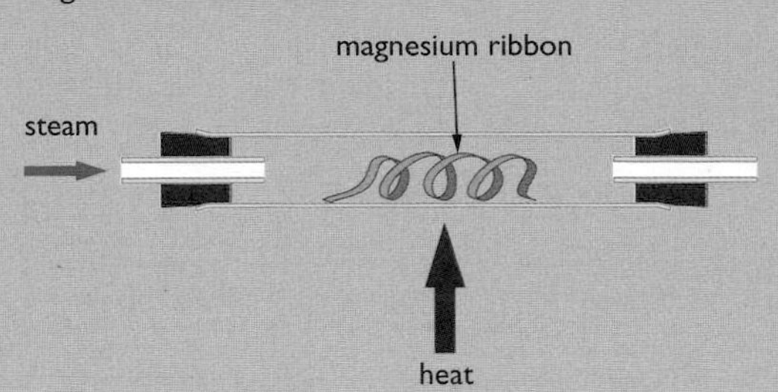


The products of this reaction are magnesium oxide and hydrogen.

(i) State the colour of magnesium and of magnesium oxide. (2)

(ii) State **two** ways in which the hydrogen could be collected. (2)

(iii) The hydrogen gas can be burned as it leaves the heated tube. Write a word equation for this reaction. (1)

*(Total 12 marks)*

**11** Use information from the table below to answer this question.

| increasing reactivity (↑) | Name of metal | Colour of solid metal | Colour of a solution of the metal(II) sulfate |
|---|---|---|---|
| | magnesium | grey | colourless |
| | zinc | grey | colourless |
| | iron | dark grey | green |
| | copper | pink-brown | blue |

a) When zinc is added to magnesium sulfate solution, no reaction occurs. State why. (1)

b) When iron filings are added to copper(II) sulfate solution, a reaction takes place.

(i) Write a chemical equation for this reaction. (2)

(ii) Describe the colour changes to both the solid and the solution during this reaction. (4)

c) When copper is added to dilute sulfuric acid, no reaction occurs. When iron is added to dilute sulfuric acid, hydrogen gas and iron(II) sulfate solution are formed. What does this show about the reactivity of hydrogen compared to the reactivity of copper and the reactivity of iron? (2)

*(Total 9 marks)*

**12** a) A student was asked to draw a diagram to show apparatus he would use to prepare carbon dioxide gas in the laboratory. This is the diagram he drew.

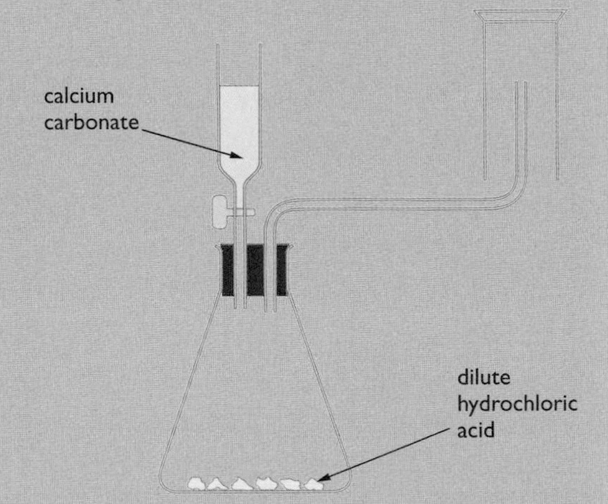


(i) State how the diagram is labelled incorrectly. (1)

(ii) Why is the method of collection of carbon dioxide unsuitable? How could the carbon dioxide be collected? (2)

(iii) Write a chemical equation, including state symbols, for the reaction that occurs in the conical flask. (3)

b) A teacher prepares a gas jar of oxygen. She then lights a piece of magnesium ribbon and places it in the gas jar. A vigorous reaction occurs. Give two observations she could make during the reaction between magnesium and oxygen. (2)

*(Total 8 marks)*

Notes

# Chapter 17: Basic definitions of terms

**Table of basic definitions of terms in organic chemistry.**

| Term | Definition |
|---|---|
| *homologous series* | a series of organic compounds that have the same general formula, similar chemical reactions and where each member differs from the next by a $-CH_2-$ group |
| *hydrocarbon* | a compound containing **only** the elements hydrogen and carbon |
| *saturated* | an organic compound in which **all** the bonds are single bonds |
| *unsaturated* | an organic compound that contains a carbon-carbon double bond |
| *general formula* | a formula that states the ratio of atoms of each element in the formula of every compound in a particular homologous series |
| *isomerism* | compounds that have the same molecular formula but different displayed formulae are said to exhibit isomerism; the different compounds are called **isomers** |

# Chapter 18: Alkanes

The alkanes are a homologous series of compounds that have the general formula $C_nH_{2n+2}$

The table shows the molecular formulae and names of the first five alkanes in the series:

| Molecular formula | Name |
|---|---|
| $CH_4$ | methane |
| $C_2H_6$ | ethane |
| $C_3H_8$ | propane |
| $C_4H_{10}$ | butane |
| $C_5H_{12}$ | pentane |

The displayed formulae of the first three members are:

*methane* *ethane* *propane*

Both butane and pentane have isomers: methylpropane is an isomer of butane; and methylbutane and dimethylpropane are isomers of pentane.

*butane* *methylpropane*

*pentane* *methylbutane* *dimethylpropane*

## Reactions of the alkanes

### Combustion

The alkanes burn when heated in air or oxygen. If there is a plentiful supply of air/oxygen, the products are carbon dioxide and water:

$$CH_4(g) + 2O_2(g) \rightarrow CO_2(g) + 2H_2O(l)$$

$$C_3H_8(g) + 5O_2(g) \rightarrow 3CO_2(g) + 4H_2O(l)$$

If there is an insufficient supply of air/oxygen, then carbon monoxide is formed:

$$CH_4(g) + 1\tfrac{1}{2}O_2(g) \rightarrow CO(g) + 2H_2O(l)$$

Carbon monoxide is poisonous. It reduces the capacity of the blood to carry oxygen.

### With bromine

Methane and bromine react together in the presence of ultra-violet (uv) radiation to form bromomethane:

$$CH_4(g) + Br_2(g) \rightarrow CH_3Br(g) + HBr(g)$$

Since one atom in methane (a hydrogen atom) has been replaced by another atom (a bromine atom) this reaction is called a **substitution reaction**.

# Chapter 19: Alkenes

The alkenes are a homologous series of compounds that have the general formula $C_nH_{2n}$

The table shows the molecular formulae and names of the first four alkenes in the series:

| Molecular formula | Name |
|---|---|
| $C_2H_4$ | ethene |
| $C_3H_6$ | propene |
| $C_4H_8$ | butene |
| $C_5H_{10}$ | pentene |

The displayed formula of ethene and propene are:

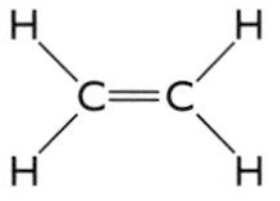

*ethene*

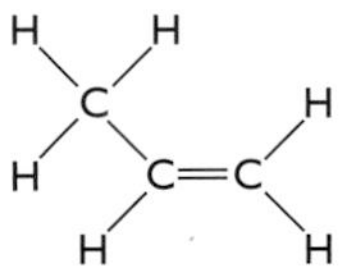

*propene*

There are in fact three isomers that are alkenes with the molecular formula $C_4H_8$:

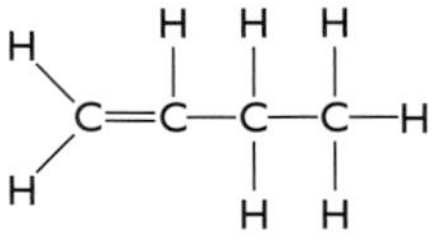

*but-1-ene*

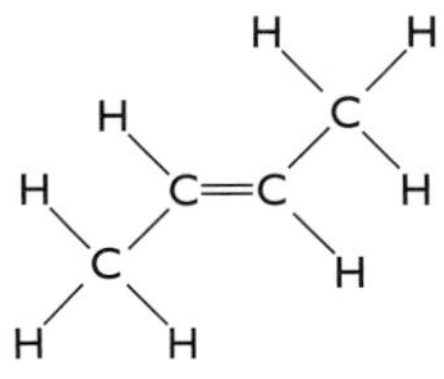

*but-2-ene*

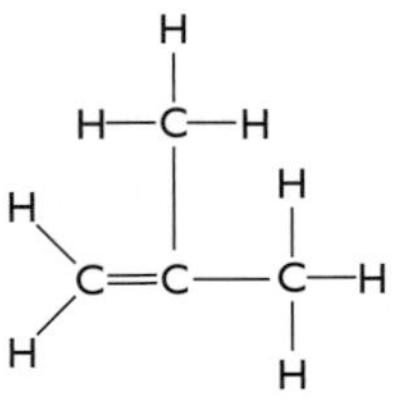

*methylpropene*

## Reactions of the alkenes

Alkenes undergo **addition reactions** with halogens. For example, a bromine molecule will add across the double bond of ethene to form 1,2-dibromoethane:

$$CH_2{=}CH_2 + Br_2 \rightarrow BrCH_2{-}CH_2Br$$

*1,2-dibromoethane*

1,2-dibromoethane is colourless, so when bromine, or bromine water, is shaken with ethene the bromine will be decolourised.

Since all alkenes contain a carbon–carbon double bond, they will all decolourise bromine.

This reaction, therefore, can be used to test for the presence of a carbon–carbon double bond (i.e. **unsaturation**) in a molecule.

**Saturated** compounds will not immediately decolourise bromine.

# **Chapter 20:** Ethanol

## Manufacture of ethanol

Ethanol is manufactured by two different processes: fermentation and direct hydration of ethane.

### 1. Fermentation

- Dissolve sugar or starch in water and add yeast.
- Leave the mixture to ferment at 25–40°C for several days in the absence of air.
- Filter off the excess yeast to obtain a dilute solution of ethanol.

  If the ethanol content in the mixture rises to around 15%, the yeast is killed. If a more concentrated solution of ethanol is required the mixture is fractionally distilled.

  Whatever the starting point, sugar or starch, the enzymes in the yeast produce glucose, $C_6H_{12}O_6$. The enzymes in yeast then convert the glucose into ethanol:

$$C_6H_{12}O_6(aq) \rightarrow 2C_2H_5OH(aq) + 2CO_2(g)$$

### 2. Direct hydration of ethene

- A mixture of ethene and steam is passed over a phosphoric acid catalyst at a temperature of 300°C and 60–70 atmospheres of pressure:

$$C_2H_4(g) + H_2O(g) \rightarrow C_2H_5OH(g)$$

- The ethanol is condensed as a liquid.
- The ethene required for this reaction is obtained from crude oil ***(see Chapter 26)***.

Section C: Organic Chemistry

## Comparing methods

The table shows the similarities and differences in the two methods of producing ethanol.

| | Fermentation | Hydration of ethene |
|---|---|---|
| raw materials | uses renewable resources, e.g. sugar cane | uses non-renewable resources - once all the crude oil is used up there will not be any more |
| type of process | batch process | continuous process |
| rate of reaction | very slow, taking several days for each batch | fast |
| quality of product | produces a dilute solution of ethanol that needs further processing if pure ethanol is required | produces pure ethanol |
| reaction conditions | low temperatures required | high temperatures and pressures required, increasing the cost |

The better method to choose will depend on many factors. For example, if a country has limited access to crude oil but has the type of climate and the area of land required to grow sugar cane, then it would be more economical to use fermentation rather than direct hydration.

Also, a dilute solution of ethanol is all that is required for some uses, for example the production of wine vinegar. In this case, fermentation would again be the better option.

However, if pure ethanol is required, for example for use as a solvent, then the continuous process would be more economical.

# Reactions of ethanol

## Combustion

Ethanol burns when heated in air or oxygen:

$$C_2H_5OH(l) + 3O_2(g) \rightarrow 2CO_2(g) + 3H_2O(l)$$

For this reason it can be used as a fuel. A fuel for cars that contains a mixture of ethanol and petrol is now available in many countries.

## Dehydration

Ethanol can be dehydrated (i.e. have the elements of water removed from it) by passing ethanol vapour over hot aluminium oxide, which acts as a catalyst for the reaction:

$$C_2H_5OH(g) \rightarrow C_2H_4(g) + H_2O(g)$$

This reaction can be used to manufacture ethene for the production of poly(ethene) ***(see Chapter 27)***. However, at present most ethene is made by cracking crude oil fractions ***(see Chapter 26)***.

## Examination Questions

**1** The molecular formulae of four organic compounds, **A**, **B**, **C** and **D**, are shown below.

**A** $CH_4$ **B** $C_2H_4$ **C** $C_2H_6$ **D** $C_3H_8$

*a)* Explain why all four compounds are hydrocarbons. *(1)*

*b)* Compounds in the same homologous series have the same general formula.

(i) State the name of the homologous series to which compound **C** belongs. *(1)*

(ii) Which one of the following formulae represents the general formula of this homologous series? *(1)*

$C_nH_{2n-2}$ $C_nH_{2n}$ $C_nH_{2n+2}$

*(iii)* State which of the compounds **A**, **B** and **D** are members of the same homologous series as **C**. *(1)*

*c)* (i) Explain the term **isomers**. *(2)*

(ii) State which, if any, of the four compounds have isomers. *(1)*

*d)* Draw a displayed formula of a molecule of **B** showing the arrangement of the bonds around the carbon atoms. *(2)*

*e)* Calculate the relative formula mass of **B** using information from the Periodic Table (p. 119). *(1)*

***(Total 10 marks)***

**2** The alkenes are a homologous series of unsaturated hydrocarbons.

*a)* (i) Indicate which **two** of the following statements about members of a homologous series are correct:

- They have similar chemical properties
- They have the same displayed formula
- They have the same general formula
- They have the same physical properties
- They have the same relative formula masses *(2)*

(ii) What is meant by the term **unsaturated**? *(1)*

*b)* Alkenes react with bromine water. Ethene is the simplest alkene.

(i) Bromine water is added to ethene. State the starting and finishing colours of the reaction mixture. *(2)*

(ii) Complete the equation by drawing the displayed formula of the product. *(2)*

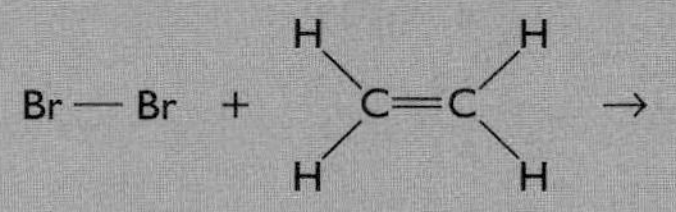

Br — Br + $H_2C{=}CH_2$ →

*c)* Isomers are compounds that have the same molecular formula but different displayed formulae.

Draw the displayed formulae of **two** isomers that have the molecular formula $C_4H_8$. *(2)*

***(Total 9 marks)***

**3** Sugar can be converted into poly(ethene) as follows:

sugar —Reaction 1→ ethanol —Reaction 2→ ethene —Reaction 3→ poly(ethene)

*a)* (i) State the type of reaction occurring in both Reaction 1 and Reaction 2. *(2)*

(ii) What type of polymerisation occurs in Reaction 3? *(1)*

*b)* State **two** conditions used in the conversion of sugar to ethanol in Reaction 1. *(2)*

*c)* Write a chemical equation for Reaction 2. *(2)*

*d)* Draw the displayed formula of ethanol. *(1)*

*e)* Many thousands of ethene molecules combine to form a poly(ethene) molecule. Draw that part of the structure of a poly(ethene) molecule that forms from **three** ethene molecules. *(2)*

*f)* Nylon is made by a different type of polymerisation. Name this type of polymerisation and describe how it is different from the type of polymerisation used to make poly(ethene). *(2)*

***(Total 12 marks)***

**4** These are the structures of six hydrocarbons.

**A** **B** 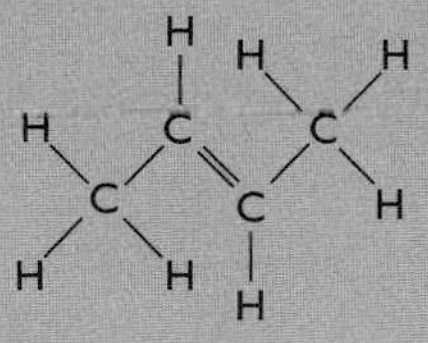 **C** 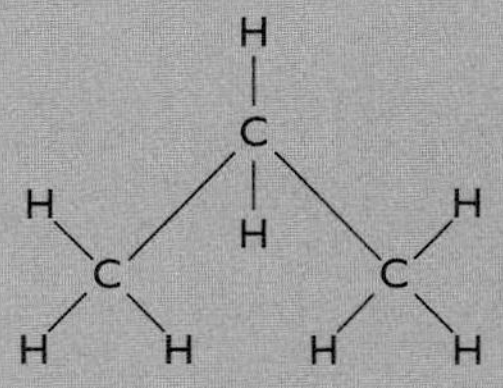 **D** 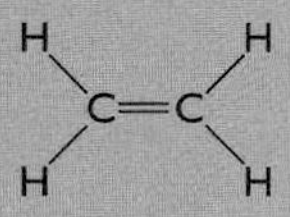 **E** 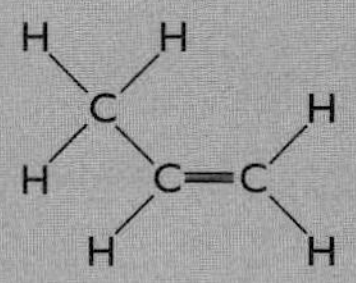 **F**

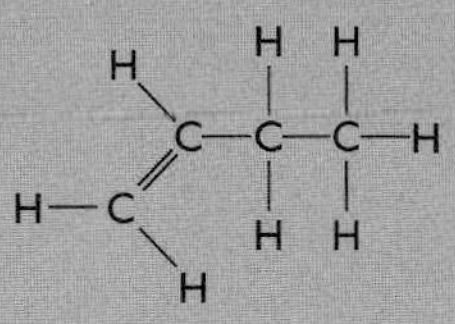

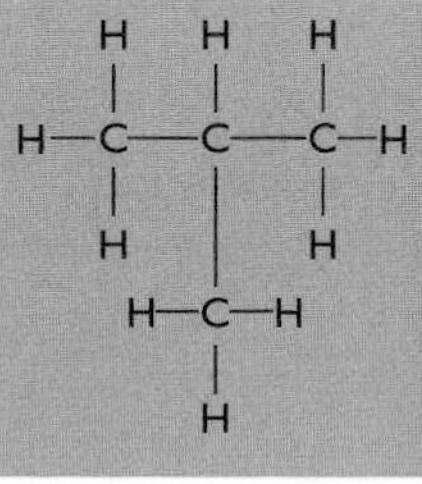

a) For hydrocarbons **A–F**, answer these questions.

(i) Which hydrocarbon is **not** an alkene. *(1)*

(ii) Which **two** hydrocarbons are isomers? *(1)*

(iii) Which structure is propene? *(1)*

b) Hydrocarbon **D** forms a polymer. Give the name of this polymer and draw a diagram to represent the structure of the polymer. *(3)*

***(Total 6 marks)***

**5** *a)* The table below shows the displayed formulae of some organic compounds

| Compound | Displayed Formula |
|---|---|
| **A** | H–C(H)(H)–C(H)(H)–H |
| **B** | H–C(H)(H)–C(H)(H)–O–H |
| **C** | H₂C=CH₂ (H, H on each C) |
| **D** | H–C(H)(H)–C(H)(H)–C(H)(H)–H |
| **E** | H₂C=CH–CH₃ (H, H on C; C=C; H and C(H)(H)(H)) |

(i) Give one reason why compound **B** is not a hydrocarbon. *(1)*

(ii) State the empirical formula of compound **A**. *(1)*

(iii) Both **A** and **D** are members of the same homologous series.

What is a homologous series? *(2)*

(iv) Draw a dot and cross diagram to show the bonding in compound **A**. *(2)*

(v) What is the name of the addition polymer formed by compound **E**? *(1)*

(vi) Draw the repeat unit of the addition polymer of compound **E**. *(2)*

(vii) Compound **E** reacts rapidly with bromine water but the addition polymer of compound **E** does not. Explain this difference in behaviour. *(2)*

*b)* Draw the displayed formulae of three isomers with molecular formula $C_4H_8$. *(3)*

***(Total 14 marks)***

**Notes**

Notes

# Chapter 21: Acids, alkalis and salts

## The pH scale

As you go through your chemistry course, you will find that the meaning of the word **acid** changes gradually. To start with, we say that an acid is a substance that has a sour taste, though tasting acids is not recommended since some of them are poisonous. The word acid often suggests a substance that burns or corrodes, but many acids are quite harmless. Indeed, much of what we eat and drink contains one or more acids.

At a more advanced level, water is best described as an amphoteric substance, i.e. having the properties of both acids and bases. However, at this level it is best to consider it to be neutral.

pH is **lower case** p and **upper case** H.

**EXAMINER'S TIP**

1. Note that the word alkaline is an adjective and the word alkali is a noun. Thus we say 'an alkaline solution' but we speak of 'an alkali' not 'an alkaline'.
2. pH values only exist when the substance is dissolved in water. Thus a completely insoluble substance cannot have a pH value. Universal indicator is used to measure pH values and this, along with other acid-alkali indicators, is described in the next section.

**EXAMINER'S TIP**

The word **indicator** often implies an acid-alkali indicator, but a substance such as anhydrous copper(II) sulfate or cobalt(II) chloride are also indicators, since they both indicate the presence of water.

The chemical 'opposite' of an acid is a **base**. Bases that are soluble in water are called **alkalis**. Alkalis are often much more harmful than acids. They should be treated with great care. For example, they can be very damaging to your eyes.

A substance that has neither acidic nor alkaline properties is described as being **neutral**. The best example of a neutral substance is pure water.

The acidity and alkalinity of aqueous solutions of substances can be measured using the pH scale.

Neutral solutions have a pH of 7, acidic solutions have values less than 7 and alkaline solutions have values greater than 7. A very low pH means a very strongly acidic solution whereas a value between 4 and 6 shows a weakly acidic solution; the lower the value, the more strongly acidic is the solution.

The opposite applies to alkalis, where very high values indicate very strongly alkaline solutions.

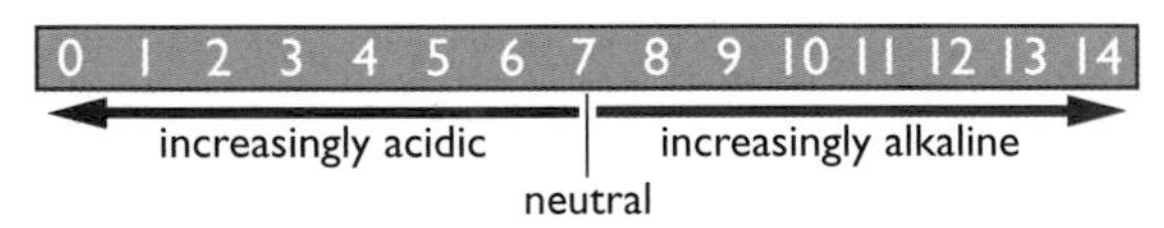


**Figure 21.1** *The pH scale.*

### Acid-alkali indicators

It has been known for thousands of years that certain dyes change colour when an acidic solution or an alkaline solution is added to them. These dyes are therefore able to **indicate** the presence of acids or alkalis in aqueous solutions.

One of the first indicators was a substance called **litmus**. Litmus is extracted from a species of lichen, a type of fungus that grows on trees. In solutions where the pH is 5 or less, litmus turns red; in solutions of pH 8 or more, litmus turns blue. In solutions of pH between 5 and 8, the litmus is a shade of purple (since red and blue mix to give purple). Hence, **'neutral'** litmus is purple. Litmus is said to have a **pH range** of 5 to 8.

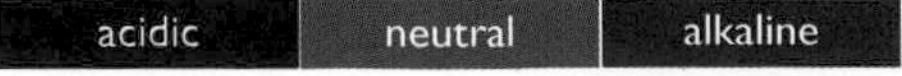


**Figure 21.2** *The colour of litmus under different conditions.*

Other indicators have different pH ranges. The chart below shows the pH ranges of three common indicators.

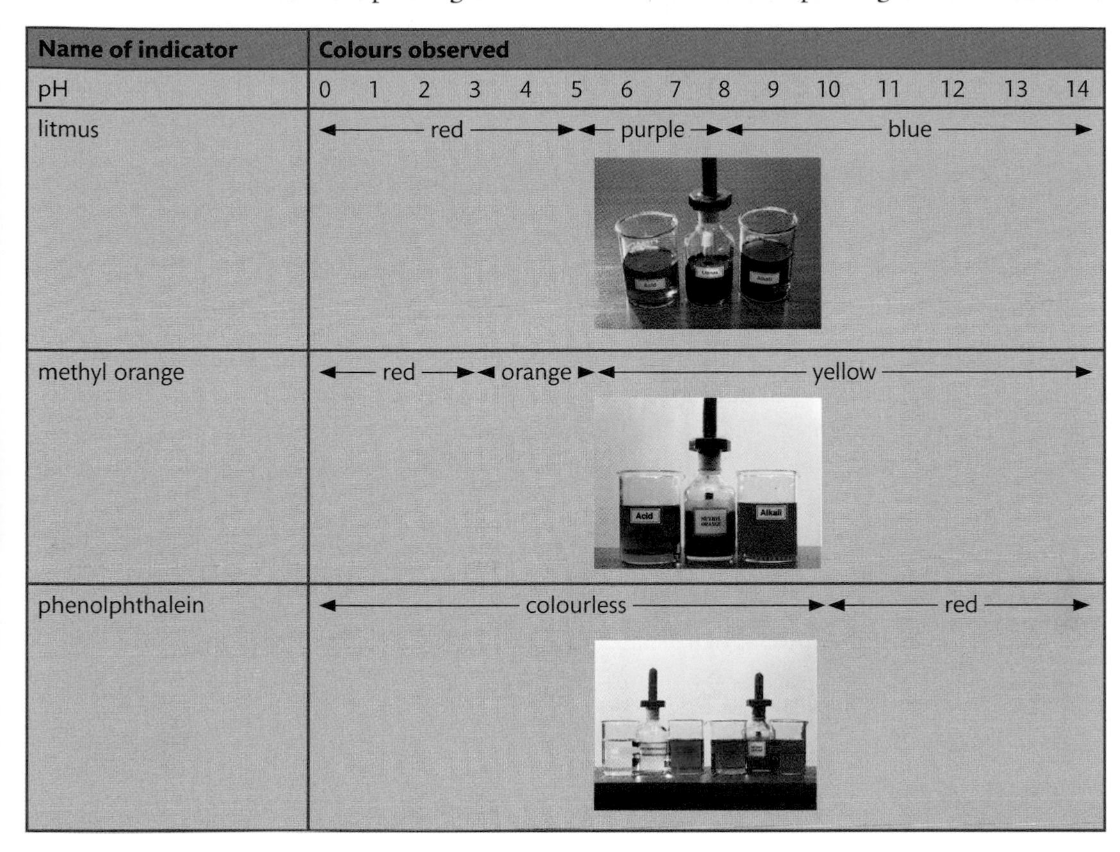

| Name of indicator | Colours observed |
|---|---|
| pH | 0 1 2 3 4 5 6 7 8 9 10 11 12 13 14 |
| litmus | ◄— red —►◄— purple —►◄— blue —► |
| methyl orange | ◄— red —►◄ orange ►◄— yellow —► |
| phenolphthalein | ◄— colourless —►◄— red —► |

## Universal indicator and measuring pH

It is easy to see that the use of any one of these indicators to prove that a solution is **definitely** acidic or alkaline has its limitations – hence the search for and development of a single '**universal indicator**'. By selecting six or seven indicators that change colour at different pH values, it is possible to make up an indicator that changes colour gradually over the whole pH range. Often the indicators are chosen so that the colours of the spectrum (red, orange, yellow, green, blue, indigo, and violet) are obtained, starting with red at the acid end (pH 0).

Universal indicator is supplied in two forms. One is a solution in ethanol and the other is paper that has been soaked in the indicator solution and then dried. The paper is often known as **pH paper**.

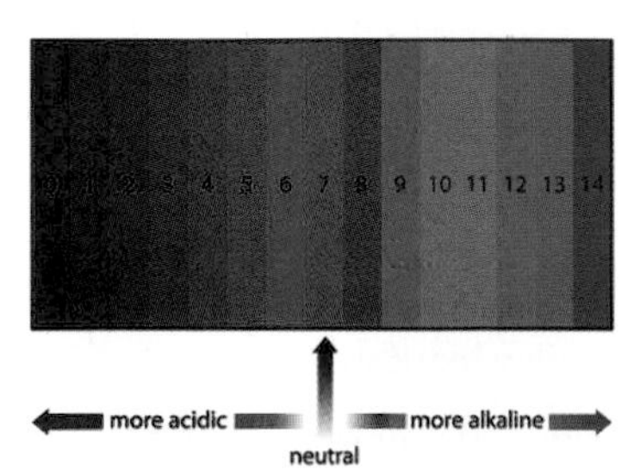


**Figure 21.3** *Colour range and equivalent pH of liquid universal indicator.*

The simplest way to find an unknown pH is to use universal indicator. The substance with the unknown pH is dissolved in water and then a drop of it is placed onto pH paper. If a solution of universal indicator is used, then one drop of it should be added to the solution of the substance under test. If the pH of a gas is required, it can be found by testing the gas with **damp** pH paper.

## Reactions of acids

The best definition of an acid at International GCSE is 'a substance that dissolves in water to produce hydrogen ions ($H^+$)'. The table below gives the name, formula and ionisation reaction for the three most common acids found in a school laboratory:

| Name of acid | Formula of acid | Ionisation reaction in water |
|---|---|---|
| hydrochloric acid | $HCl$ | $HCl(aq) \rightarrow H^+(aq) + Cl^-(aq)$ |
| nitric acid | $HNO_3$ | $HNO_3(aq) \rightarrow H^+(aq) + NO_3^-(aq)$ |
| sulfuric acid | $H_2SO_4$ | $H_2SO_4(aq) \rightarrow 2H^+(aq) + SO_4^{2-}(aq)$ |

When the hydrogen ions in an acid are replaced by either a metal ion or by an ammonium ion ($NH_4^+$), the compound formed is called a **salt**. The name of the salt formed is determined by both the parent acid and the ion that replaces the hydrogen ion. The first half of the name is derived from the ion that replaces the hydrogen ion, whereas the second half of the name is derived from the parent acid. Examples are shown in the table:

| Parent acid | Salt formed | Name of a typical salt |
|---|---|---|
| hydrochloric acid | chloride | sodium chloride |
| nitric acid | nitrate | magnesium nitrate |
| sulfuric acid | sulfate | ammonium sulfate |

There are a number of ways to replace the hydrogen ions in an acid by a metal ion. These include reacting a dilute acid with a:

- metal
- base (i.e. a metal oxide or metal hydroxide)
- metal carbonate

### *Reacting dilute acids with metals*

Nitric acid reacts in a very strange way with most metals and hence is never used to make salts by direct reaction with metals.

Only those metals **above** hydrogen in the reactivity series ***(see Chapter 15)*** will react directly with dilute acids.

Dilute hydrochloric acid and dilute sulfuric acid react with metals to form a salt and hydrogen ***(see Chapter 15)***. The table below gives details of the reactions between several combinations of metals and dilute acids:

| | Names of products | Equation for reaction |
|---|---|---|
| magnesium + dil. hydrochloric acid | magnesium chloride and hydrogen | $Mg(s) + 2HCl(aq) \rightarrow MgCl_2(aq) + H_2(g)$ |
| zinc + dil. sulfuric acid | zinc sulfate and hydrogen | $Zn(s) + H_2SO_4(aq) \rightarrow ZnSO_4(aq) + H_2(g)$ |
| iron + dil. sulfuric acid | iron(II) sulfate and hydrogen | $Fe(s) + H_2SO_4(aq) \rightarrow FeSO_4(aq) + H_2(g)$ |
| aluminium + dil. hydrochloric acid | aluminium chloride and hydrogen | $2Al(s) + 6HCl(aq) \rightarrow 2AlCl_3(aq) + 3H_2(g)$ |

### *Reacting dilute acids with bases*

All metal oxides and metal hydroxides are capable of acting as bases. Bases neutralise acids to form a salt. The following table gives details of the reactions between several combinations of bases and dilute acids:

| | Names of products | Equation for reaction |
|---|---|---|
| magnesium oxide + dil. hydrochloric acid | magnesium chloride and water | $MgO(s) + 2HCl(aq) \rightarrow MgCl_2(aq) + H_2O(l)$ |
| zinc hydroxide + dil. nitric acid | zinc nitrate and water | $Zn(OH)_2(s) + 2HNO_3(aq) \rightarrow Zn(NO_3)_2(aq) + 2H_2O(l)$ |
| copper(II) oxide + dil. sulfuric acid | copper(II) sulfate and water | $CuO(s) + H_2SO_4(aq) \rightarrow CuSO_4(aq) + H_2O(l)$ |
| sodium hydroxide + dil. hydrochloric acid | sodium chloride and water | $NaOH(aq) + HCl(aq) \rightarrow NaCl(aq) + H_2O(l)$ |

### *Reacting dilute acids with metal carbonates*

Dilute acids react with metal carbonates to form salts, carbon dioxide and water. The table below gives details of the reactions between several combinations of carbonates and dilute acids:

| | Names of products | Equation for reaction |
|---|---|---|
| magnesium carbonate + dil. hydrochloric acid | magnesium chloride, carbon dioxide and water | $MgCO_3(s) + 2HCl(aq) \rightarrow MgCl_2(aq) + CO_2(g) + H_2O(l)$ |
| zinc carbonate + dil. nitric acid | zinc nitrate, carbon dioxide and water | $ZnCO_3(s) + 2HNO_3(aq) \rightarrow Zn(NO_3)_2(aq) + CO_2(g) + H_2O(l)$ |
| copper(II) carbonate + dil. sulfuric acid | copper(II) sulfate, carbon dioxide and water | $CuCO_3(s) + H_2SO_4(aq) \rightarrow CuSO_4(aq) + CO_2(g) + H_2O(l)$ |
| sodium carbonate + dil. hydrochloric acid | sodium chloride, carbon dioxide and water | $Na_2CO_3(s) + 2HCl(aq) \rightarrow 2NaCl(aq) + CO_2(g) + H_2O(l)$ |

## Solubility of salts

The method used to prepare a salt depends on whether the salt is soluble or insoluble in water. Hence it is necessary to know the general rules that govern the solubility of the salts of the common metals. These are outlined in the table below:

| Soluble salts | Insoluble salts |
|---|---|
| all common sodium, potassium and ammonium salts | |
| all nitrates | |
| all common chlorides (except silver and lead(II) chloride) | silver chloride and lead(II) chloride |
| all common sulfates (except barium, calcium and lead(II) sulfate) | barium sulfate, calcium sulfate and lead(II) sulfate |
| sodium, potassium and ammonium carbonates | all other common carbonates |

### Preparation of soluble salts

The three common methods for preparing soluble salts are:

- dilute acid + base (i.e. metal oxide or metal hydroxide)
- dilute acid + metal carbonate
- dilute acid + metal

### *Dilute acid + base*

The practical technique used depends on whether the base is insoluble or soluble in water.

The only common bases that are very soluble in water are sodium hydroxide and potassium hydroxide.

Some salts, such as copper(II) sulfate, are much more soluble in hot water than in cold water. In these cases, crystals can be obtained very quickly by partially evaporating the dilute solution of the salt by boiling until a hot saturated solution is formed. When this solution is allowed to cool, crystals will form. These crystals can then be removed by filtration and dried on filter paper.

1. Acid + insoluble base
   - Put some dilute acid into a beaker and heat it using a Bunsen burner flame. Do not let it boil.
   - Add the insoluble base, a little at a time, to the warm dilute acid and stir until the base is in excess (i.e. until the base stops disappearing and a suspension of the base forms in the acid).
   - Filter the mixture into an evaporating basin to remove the excess base.
   - Leave the filtrate (i.e. the dilute solution of the salt) in a warm place in the laboratory so that the water evaporates and crystals form.
   - Remove the crystals and dry them on filter paper.

If the alkali is added to the acid the colour changes of the indicator are reversed.

2. Acid + soluble base (alkali)
   - Put an aqueous solution of the alkali into a conical flask and add a suitable indicator (e.g. litmus or methyl orange).
   - Add dilute acid from a burette until the indicator just changes colour.
   - Add powdered charcoal and shake the mixture to remove the colour of the indicator.
   - Filter to remove the charcoal and then obtain crystals from the filtrate in the usual manner.

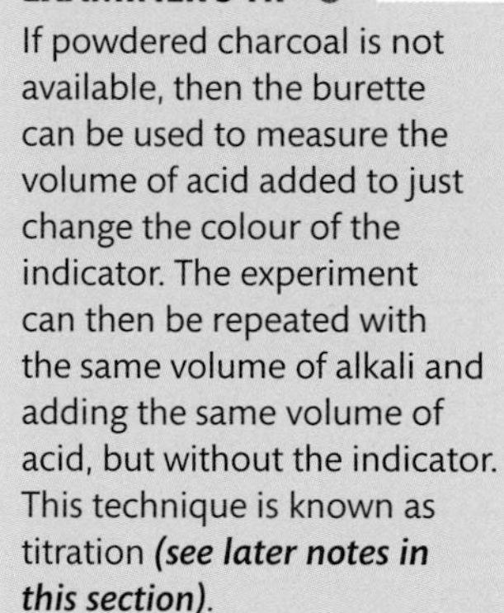
**EXAMINER'S TIP**

If powdered charcoal is not available, then the burette can be used to measure the volume of acid added to just change the colour of the indicator. The experiment can then be repeated with the same volume of alkali and adding the same volume of acid, but without the indicator. This technique is known as titration ***(see later notes in this section)***.

### *Dilute acid + metal carbonate*

Most metal carbonates are insoluble in water and hence the method for acid + insoluble base can be followed. A good indication this time of complete neutralisation of the acid is to wait until no more bubbles of gas are formed.

Sodium and potassium carbonates are soluble in water and, therefore, if a salt of either sodium or potassium is required, the procedure for acid + soluble base must be followed. The only indicator suitable for obtaining the neutralisation point is methyl orange, which changes colour from yellow in a solution of the carbonate, to red in acid.

### *Dilute acid + metal*

The only metals commonly used are magnesium, iron and zinc. Salts of nitric acid are **never** made this way.

Since all three metals are insoluble in water, and do not react with water, the procedure for acid + insoluble base is adopted. Again, since hydrogen gas is evolved, the best sign that all the acid has reacted is a lack of effervescence when more metal is added.

## Preparation of insoluble salts

Insoluble salts are made by **precipitation** reactions.

A **precipitate** is an insoluble solid that is made by a chemical reaction that takes place in aqueous solution.

A **precipitation reaction** is a reaction that produces a precipitate.

To make an insoluble salt it is necessary to mix together two separate aqueous solutions. One solution must contain the required positive ion and the other solution the required negative ion. For example, to make silver chloride, mix together silver nitrate solution and sodium chloride solution:

$$AgNO_3(aq) + NaCl(aq) \rightarrow AgCl(s) + NaNO_3(aq)$$

The silver nitrate provides the silver ions, $Ag^+$

The sodium chloride provides the chloride ions, $Cl^-$

The precipitate of silver chloride is then removed by filtration, washed with a little distilled water (to remove traces of the potassium nitrate solution) and left to dry in a warm place.

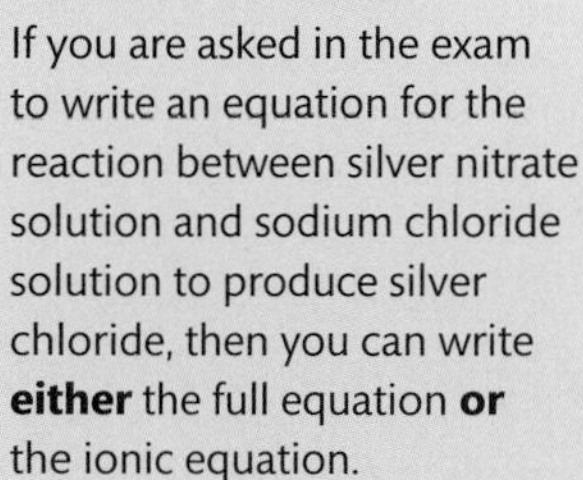

**EXAMINER'S TIP**

If you are asked in the exam to write an equation for the reaction between silver nitrate solution and sodium chloride solution to produce silver chloride, then you can write **either** the full equation **or** the ionic equation.

There is only **ONE** equation to learn.

The ionic equation for the above reaction is:

$$Ag^+(aq) + Cl^-(aq) \rightarrow AgCl(s)$$

The advantage of the ionic equation is that it can be used for the reaction between any solution containing silver ions and any solution containing chloride ions. Hence, the same equation applies to the reaction between silver sulfate solution and magnesium chloride.

The table below gives some examples of solutions that can be mixed together to make insoluble salts:

| Solution providing the positive ion | Solution providing the negative ion | Name of insoluble salt formed | Ionic equation for the reaction |
|---|---|---|---|
| barium chloride | potassium sulfate | barium sulfate | $Ba^{2+}(aq) + SO_4^{2-}(aq) \rightarrow BaSO_4(s)$ |
| lead(II) nitrate | zinc chloride | lead(II) chloride | $Pb^{2+}(aq) + 2Cl^-(aq) \rightarrow PbCl_2(s)$ |
| magnesium sulfate | sodium carbonate | magnesium carbonate | $Mg^{2+}(aq) + CO_3^{2-}(aq) \rightarrow MgCO_3(s)$ |
| silver nitrate | barium iodide | silver iodide | $Ag^+(aq) + I^-(aq) \rightarrow AgI(s)$ |

## Making ammonium salts

An ammonium salt is made whenever ammonia, $NH_3$, reacts with an acid. Examples include:

$$NH_3(aq) + HCl(aq) \rightarrow NH_4Cl(aq)$$

$$NH_3(aq) + HNO_3(aq) \rightarrow NH_4NO_3(aq)$$

$$2NH_3(aq) + H_2SO_4(aq) \rightarrow (NH_4)_2SO_4(aq)$$

These are all examples of a neutralisation reaction. In order to decide when the reaction is complete, an indicator has to be used. So, for example, if you were reacting ammonia with hydrochloric acid, you would add litmus to the acid and then, using a burette, slowly add ammonia solution to the acid until the litmus changes colour. You could then repeat the reaction using the same volumes of solutions, but without the indicator. The final solution can then be left to crystallise in the usual way.

## Acid–alkali titrations

A **titration** is a method of finding out exactly the volume of one solution that is required to react with a given volume of another solution.

Titrations are commonly used to find the volume of acid required to react exactly with a given volume of an alkali. Such titrations are called **acid–alkali** titrations.

The method of carrying out an acid–alkali titration is as follows:

- using a pipette, put 25.0 $cm^3$ of the alkali solution into a conical flask
- add a few drops of an indicator, such as methyl orange
- put the acid into a burette and note the initial reading
- add the acid to the alkali until the indicator just changes colour (e.g. methyl orange turns from yellow to orange – ***see p. 69***)
- note the final reading of acid in the burette
- subtract the initial reading from the final reading to obtain the volume of acid added. This is the volume required to neutralise the 25.0 $cm^3$ of the alkali.

# Chapter 22: Energetics

## Exothermic and endothermic reactions

Most chemical reactions are accompanied by a change in heat energy. There are two types of heat energy change in reactions – exothermic and endothermic. The table outlines the essential features of these reactions:

| | Exothermic reaction | Endothermic reaction |
|---|---|---|
| **Definition** | energy is given out | energy is taken in |
| **Example** | the reaction between sulfuric acid and magnesium | the reaction between sodium hydrogen carbonate and citric acid |
| **Measuring the energy change** | 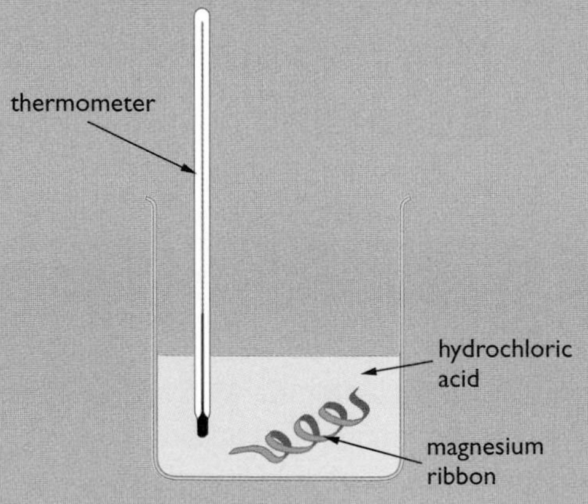 thermometer; hydrochloric acid; magnesium ribbon<br>There is a rise in temperature | 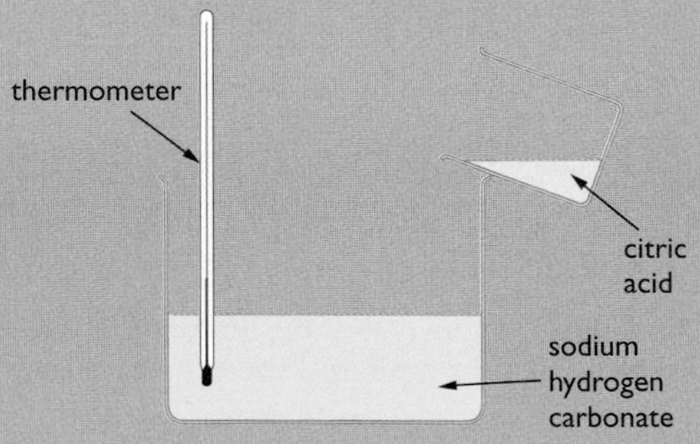 thermometer; citric acid; sodium hydrogen carbonate<br>There is a fall in temperature |

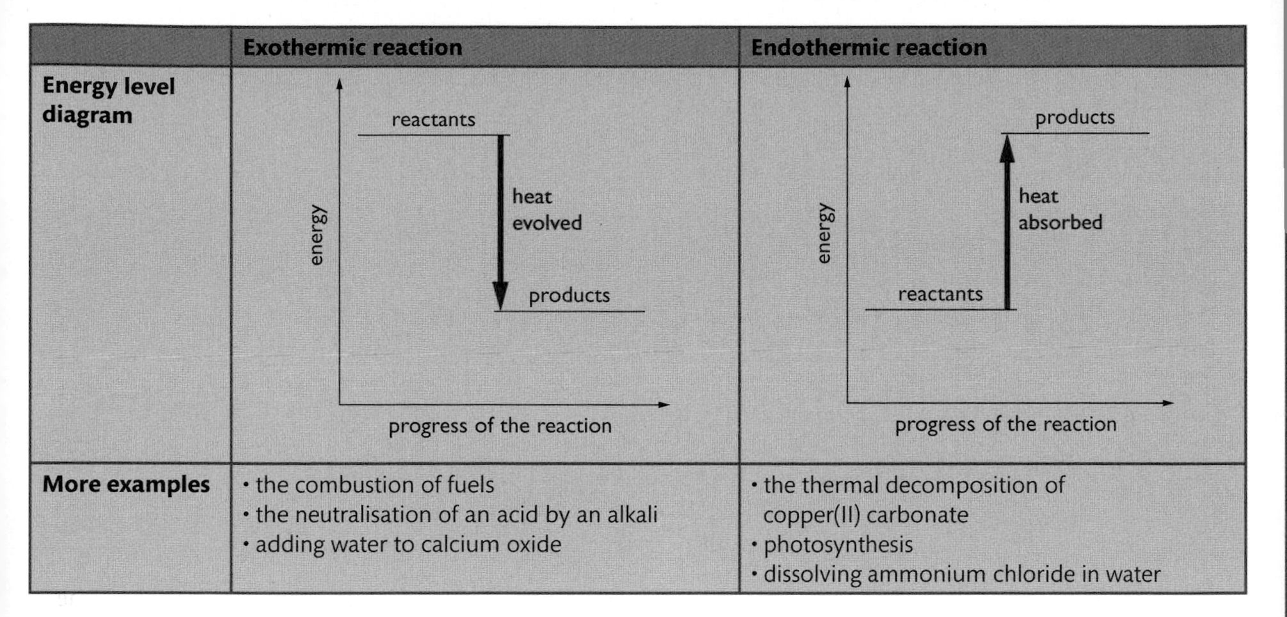


| | Exothermic reaction | Endothermic reaction |
|---|---|---|
| **Energy level diagram** | | |
| **More examples** | • the combustion of fuels<br>• the neutralisation of an acid by an alkali<br>• adding water to calcium oxide | • the thermal decomposition of copper(II) carbonate<br>• photosynthesis<br>• dissolving ammonium chloride in water |

The amount of heat energy **change** is given the symbol ΔH. ***(See worked example below for a more accurate description of ΔH.)***

For an **exothermic** reaction, ΔH is given a **negative** sign.

For an **endothermic reaction**, ΔH is given a **positive** sign.

## Experimental work

### *Measuring the heat energy change during reactions in solution*

Example: the reaction between magnesium and hydrochloric acid

- using a measuring cylinder, place 50 $cm^3$ of 0.1 $mol/dm^3$ hydrochloric acid into a polystyrene cup supported in a beaker
- measure and record the temperature of the acid
- tip approximately 0.15 g of magnesium powder (an excess) into the acid and stir the mixture
- measure and record the highest temperature reached by the mixture

Calculations:

- calculate the rise in temperature
- calculate the heat given out during the reaction by using the equation

**heat given out =**
**mass of solution × specific heat capacity of solution × temperature rise**

The specific heat capacity of the solution is the amount of heat needed to raise the temperature of one gram (1g) of solution by one degree Celsius (1°C).

This equation is sometimes abbreviated to $Q = mc\Delta T$, where Q = heat given out; m = mass of solution; c = specific heat capacity; ΔT = temperature rise

The specific heat capacity of water is 4.2 (joules per gram per degree Celsius). For dilute solutions of hydrochloric acid the value is the same. We can also assume that the density of the acid is the same as that of water (1 g/cm$^3$), so the mass of solution will be 50 g.

**Worked Example with Specimen Results**

Temperature rise = 10.0°C

Moles of acid = $\frac{50}{1000} \times 0.1 = 0.005$

Heat given out = $50 \times 4.2 \times 10.0$ J = 2100 J = 2.10 kJ

0.005 mol of acid produce 2.10 kJ

∴ 1 mol of acid produces = $\frac{2.10}{0.005}$ = 420 kJ

∴ **ΔH = −420 kJ/mol**

For reactions carried out at **constant pressure**, like the example above, the heat energy change is known as the **enthalpy** change. The enthalpy change per mole is called the **molar** enthalpy change and is given the symbol ΔH.

The value of ΔH is negative since the reaction is exothermic.

Polystyrene is a very good insulator so does not absorb much heat energy. To further reduce the losses you could put a lid on top of the cup during the reaction.

**EXAMINER'S TIP**

The volume of liquid added could be more accurately measured using a burette or a pipette, rather than a measuring cylinder. However, the errors involved in the experiment are such that the increased accuracy of measurement would make no difference to the overall accuracy of the experiment.

So, for the reaction between magnesium and dilute hydrochloric acid, ΔH = −420 kJ/mol.

The calculation above can be performed for any reaction that is taking place in aqueous solution. This includes neutralisation reactions between acids and alkalis, dissolving solids in water and also displacement reactions, where a metal is added to an aqueous solution of the salt of a less reactive metal.

In each case the same assumptions will be made. These are:

- the specific heat capacity of the final solution is the same as that for water, 4.2 J/g/°C
- the density of the final solution is 1 g/cm$^3$
- there is very little heat lost to the polystyrene cup and the surroundings, including the thermometer.

### *Measuring the heat energy change during combustion reactions*

The most common experiments are those involving liquid fuels such as an alcohol. The liquid is burnt and the heat generated is used to heat up some water. The temperature rise of the water is measured and the heat generated is calculated in the same way as the previous worked example.

Procedure:

- using a measuring cylinder, put 100 cm$^3$ of water into a copper can
- measure and record the initial temperature of the water
- fill the spirit burner with alcohol and measure and record its mass
- place the burner under the copper can and light the wick

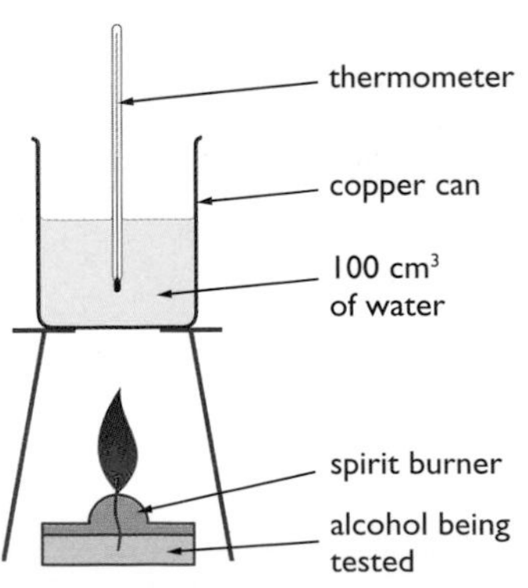


**Figure 22.1** *The apparatus used for heat energy change experiments.*

- stir the water constantly with the thermometer and continue heating until the temperature rises by about 20–30°C. Blow out the flame
- measure and record the highest temperature of the water
- measure and record the final mass of burner and remaining alcohol

Calculations:

- calculate the rise in temperature of the water
- calculate the mass of alcohol burnt
- calculate the heat given out by the reaction using $Q = mc\Delta T$
- calculate the molar enthalpy of reaction (i.e. the amount of heat energy released when one mole of the alcohol is burnt)

**Worked Example with Specimen Results**

Temperature rise = 24.5°C

Mass of ethanol burnt = 0.46 g

Heat given out = 100 × 4.2 × 24.5 J = 10290 J = 10.29 kJ

$M_r$ of ethanol = 46 ∴ 1 mol of ethanol = 46 g

0.46 g of ethanol produce 10.29 kJ

∴ 46 g of ethanol produce $\frac{10.29}{0.46} \times 46$ kJ = 1029 kJ

∴ **ΔH = −1029 kJ/mol**

The accepted value of ΔH for ethanol is −1370 kJ/mol. The value produced in the above experiment (−1029 kJ/mol) is far too low for various reasons. There are a number of sources of error – in particular, large amounts of heat losses.

The calculation assumes that all of the heat energy generated from burning the ethanol is transferred to the water. This is not the case. For example, heat energy is being used in heating the air surrounding the flame; heat energy is lost from the water by convection; heat energy is being used to raise the temperature of the copper can.

The errors in measurement (i.e. reading temperatures, measuring masses, etc.) are negligible compared with the heat losses. Hence, using a burette or pipette to measure the volume of water will not improve significantly the overall accuracy of the experiment.

## Calculations involving bond energies

The energy required to break a bond is known as the **bond energy**. It is usually measured in kilojoules per mole of bonds (kJ/mol).

The table gives the bond energies for some bonds that are commonly met at International GCSE:

| **Bond** | C–H | H–H | O=O | H–Cl | Cl–Cl | C=O | O–H |
|---|---|---|---|---|---|---|---|
| **Bond energy in kJ/mol** | 412 | 436 | 496 | 432 | 242 | 743 | 463 |

Breaking bonds takes in energy (i.e. is an endothermic process).

Making bonds gives out energy (i.e. is an exothermic process).

An approximate value for the overall enthalpy change for some chemical reactions can be calculated by considering how much energy is required to break the bonds in the reactants and comparing it with the amount of energy released when the bonds in the products are formed.

The calculation should be performed in three steps:

**Step 1:** calculate the sum of the energies for the bonds broken, $\Sigma$ (bonds broken)

**Step 2:** calculate the sum of the energies for the bonds made, $\Sigma$ (bonds made)

**Step 3:** calculate $\Delta H$ using the formula $\Delta H = \Sigma$ (bonds broken) $- \Sigma$ (bonds made)

**Worked Example 1**

The reaction between hydrogen and chlorine.

$H_2(g) + Cl_2(g) \rightarrow 2HCl(g)$ **or** $H{-}H(g) + Cl{-}Cl(g) \rightarrow 2H{-}Cl(g)$

$\Sigma$ (bonds broken) $= (H{-}H) + (Cl{-}Cl) = 436 + 242 = 678$ kJ

$\Sigma$ (bonds made $= 2(H{-}Cl) = 2 \times 432 = 864$ kJ

**$\Delta H = 678 - 864 = -186$ kJ/mol**

**Worked Example 2**

The reaction between methane and oxygen.

$CH_4(g) + 2O_2(g) \rightarrow CO_2(g) + 2H_2O(l)$

$\Sigma$ (bonds broken) $= 4(C{-}H) + 2(O{=}O) = (4 \times 412) + (2 \times 496) = 2640$ kJ

$\Sigma$ (bonds made) $= 2(C{=}O) + 4(O{-}H) = (2 \times 743) + (4 \times 463) = 3338$ kJ

**$\Delta H = 2640 - 3338 = -698$ kJ/mol**

# **Chapter 23:** Rates of reaction

## What is meant by rate of reaction?

Rate is a measure of how fast a reaction occurs.

**The best definition of rate of reaction is:**

$$\text{Rate} = \frac{\text{change of concentration of reactant}}{\text{time}}$$

This definition emphasises that the rate of reaction depends on **concentration** and not on **amount** of reactant.

### The collision theory of reaction rates

The collision theory makes the following assumptions concerning chemical reactions:

- For a reaction to take place, the particles of the reactants must collide with one another.
- The colliding particles must have sufficient energy to react. This energy is known as the **activation energy**. Collisions that have the required activation energy are called **successful** collisions.

- To increase the rate of a chemical reaction, it is necessary to increase the frequency of successful collisions. That is, more successful collisions need to take place every second.

## Factors that affect the rate of a reaction

The three factors that affect the rate of a chemical reaction are:

- the concentration of a reactant
- the temperature at which the reaction takes place
- the state of division of a solid reactant.

### *Concentration*

Increasing the concentration of a reactant increases the number of particles of reactant in a given volume and hence the reacting particles will collide more often. Hence, there will be more successful collisions per second.

Decreasing the concentration will have the opposite effect.

Increasing the pressure of a gaseous reactant is the same as increasing its concentration.

### *Temperature*

Increasing the temperature increases the average kinetic energy of the reactant particles and therefore more of the collisions that take place will have the necessary activation energy to react. Hence, there will be more successful collisions per second.

Decreasing the temperature will have the opposite effect.

### *State of division*

The smaller the pieces of solid, the larger the overall surface area. This means that there will be more particles of the solid exposed to the other reactant. Hence, there will be more successful collisions per second.

Increasing the size of the pieces of solid will have the opposite effect.

Increasing the surface area of a solid has the most dramatic effect when the solid is ground up into a powder. When an iron nail is heated in a blue Bunsen flame, the iron just glows red. When iron powder (iron filings) is sprinkled into the same Bunsen flame they immediately burn and produced yellow-orange sparks.

In flour mills, the air can fill with fine flour dust which has a very large surface area. A spark can cause the flour to catch fire and explode. The same problem occurred in coal mines when the air filled with very fine coal dust.

### *Summary*

| Change | Effect on rate of reaction | Explanation |
|---|---|---|
| Increase in concentration | Increase in rate | More particles in a given volume. Particles collide more often, therefore more successful collisions per second |
| Increase in temperature | Increase in rate | Particles have more energy. More collisions have the required activation energy, therefore more successful collisions per second |
| Increase in surface area | Increase in rate | More particles in contact with the other reactant. Particles collide more often, therefore more successful collisions per second |

> **EXAMINER'S TIP**
> Some textbooks state that a catalyst takes no part in the reaction. This is not the case. Some catalysts actually change chemically during the reaction but are then changed back to their original state after the reaction is complete.

## Effect of catalysts on rate of reaction

A catalyst is a substance that increases the rate of a chemical reaction but is chemically unchanged at the end of the reaction.

Catalysts work by providing an **alternative pathway** for the reaction. This alternative pathway has a lower activation energy than the original pathway. This means that more of the collisions taking place will have the necessary activation energy. Hence, there will be more successful collisions per second.

> **EXAMINER'S TIP**
> It is very important how you phrase your answer when explaining how a catalyst works. You must not say that the catalyst lowers the activation energy of a reaction. The original pathway for the reaction is still available, even when the catalyst is added. If collisions between particles have enough energy they can still follow this pathway. The main point of the catalyst is that it provides an **alternative** pathway.

## Energy profile diagrams

The effect of a catalyst is best shown on energy profile diagrams such as the ones shown below:

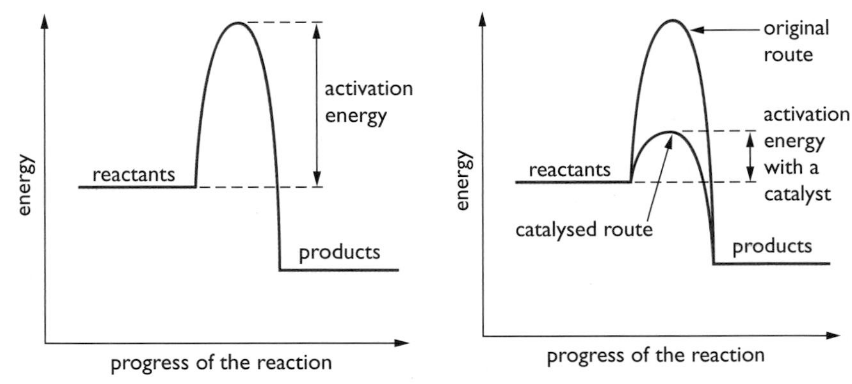


**Figure 23.1** *Activation energy of a reaction without and with a catalyst.*

Catalysts are very important in industrial reactions since they allow lower temperatures and/or pressures to be used (see Section E for examples). This saves money.

# Chapter 24: Equilibria

## Reactions that 'go both ways'

**Figure 24.1** *Heating hydrated copper(II) sulfate crystals.*

> Anhydrous means 'without water'.

### Heating copper(II) sulfate crystals, $CuSO_4.5H_2O$

When copper(II) sulfate crystals are heated in a test tube, the blue crystals turn into a white powder and a clear, colourless liquid (water) collects at the top of the test tube. The form of copper(II) sulfate in the crystals is known as **hydrated** copper(II) sulfate because it contains **water of crystallisation**. When hydrated copper(II) sulfate is heated it loses its water of crystallisation and turns into **anhydrous** copper(II) sulfate, $CuSO_4$. The equation for the reaction is:

$$\underset{\text{(blue)}}{CuSO_4.5H_2O(s)} \rightarrow \underset{\text{(white)}}{CuSO_4(s)} + 5H_2O(l)$$

When water is added to anhydrous copper(II) sulfate, there is a colour change from white to blue as it turns back into hydrated copper(II) sulfate. The equation for the reaction is:

$$CuSO_4(s) + 5H_2O(l) \rightarrow CuSO_4.5H_2O(s)$$

These two reactions can be represented by a single equation linking them both:

$$CuSO_4.5H_2O(s) \rightleftharpoons CuSO_4(s) + 5H_2O(l)$$

The $\rightleftharpoons$ sign indicates that the reaction occurs in both directions. Such reactions are said to be **reversible**.

**Figure 24.2** *Adding water to anhydrous copper(II) sulfate.*

The reaction from left to right is usually referred to as the **forward** reaction. Conversely, the reaction from right to left is referred to as the **backward** reaction.

## Heating ammonium chloride, $NH_4Cl$

Another reversible reaction takes place when ammonium chloride is heated.

If ammonium chloride is heated in a test tube, the white crystals disappear from the bottom of the tube and reappear further up. In between there is a colourless gas. Heating ammonium chloride decomposes it into the colourless gases ammonia, $NH_3$, and hydrogen chloride, $HCl$:

$$\underset{\text{(white solid)}}{NH_4Cl(s)} \rightarrow \underset{\text{(colourless gases)}}{NH_3(g) + HCl(g)}$$

As the two colourless gases rise up the tube they cool and reform ammonium chloride:

$$NH_3(g) + HCl(g) \rightarrow NH_4Cl(s)$$

Once again, the reaction is reversible and can be represented by a single equation:

$$NH_4Cl(s) \rightleftharpoons NH_3(g) + HCl(g)$$

**Figure 24.3** *Heating ammonium chloride.*

## Dynamic equilibrium

If a reversible reaction is carried out in a closed reaction container, then it is possible for the reaction to reach a position of dynamic equilibrium.

A **closed** container is one from which no reactants or products can escape.

The reaction between hydrogen gas and nitrogen gas to make ammonia gas is reversible:

$$3H_2(g) + N_2(g) \rightleftharpoons 2NH_3(g)$$

If the reaction is carried out in a closed container, then the ammonia produced cannot escape. However, since the reaction is reversible, it will start to decompose to reform hydrogen and nitrogen. When only hydrogen and nitrogen are present, i.e. at the beginning of the reaction, the rate of the forward reaction (i.e. the reaction to form ammonia) is at its highest, since the concentrations of hydrogen and nitrogen are at their highest. The rate of the backward reaction at this stage will be zero, since there is no ammonia present.

As the reaction proceeds, the concentrations of hydrogen and nitrogen gradually decrease. Hence the rate of the forward reaction will decrease. However, the concentration of ammonia is gradually increasing and hence the rate of the backward reaction will increase. Since the two reactions are interlinked, and none of the gases can escape, the rate of the forward reaction and the rate of the backward reaction will eventually become equal. This can easily be seen by using a graph of rate against time (below left):

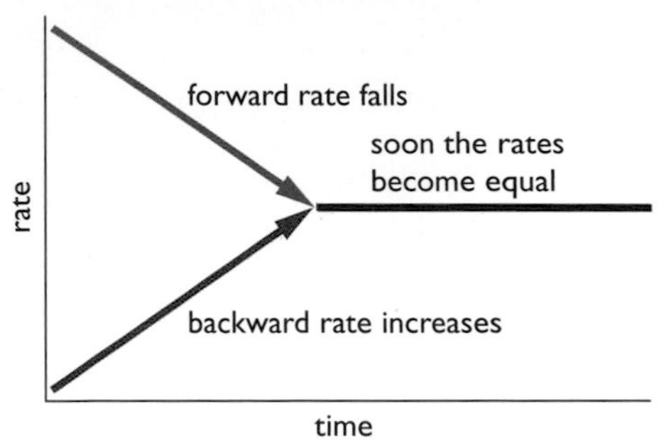


**Figure 24.4** *The rates of the forward and backward reactions become equal.*

It looks from outside as if the reaction has stopped. This is not the case. Both reactions are still occurring, but their rates are the same. When the rate of the forward reaction is equal to the rate of the backward reaction, the reaction is said to be in **dynamic equilibrium**.

A reaction that is in dynamic equilibrium has the following features:

- the rate of the forward reaction is equal to the rate of the backward reaction
- the amounts of reactants and products remain constant
- any property, such as colour or pressure, which depends on the amounts of the reactants and products that are present remains constant.

## *Controlling the equilibrium*

Here, again, is the equation for making ammonia:

$$3H_2(g) + N_2(g) \rightleftharpoons 2NH_3(g) \quad \Delta H = -92 \text{ kJ/mol}$$

This time, the value for the enthalpy change, ΔH, has been given. The value quoted for ΔH in a reversible reaction is **always** for the **forward** reaction. This means that the forward reaction is exothermic, while the backward reaction is endothermic.

By changing the pressure and temperature it is possible to **shift the position of equilibrium** of the reaction. (*Shift* means *move*.)

If the equilibrium is altered so that more ammonia is present, and less hydrogen and nitrogen, then we have shifted the position of equilibrium to the **right**.

Conversely, if the equilibrium is altered so that less ammonia is present, and more hydrogen and nitrogen, then we have shifted the position of equilibrium to the **left**.

*equilibrium shifts to the right* →

$$3H_2(g) + N_2(g) \rightleftharpoons 2NH_3(g)$$

← *equilibrium shifts to the left*

In order to predict the effect on the position of equilibrium it is necessary to know which side of the equation has the fewer number of molecules of gas and which reaction is exothermic.

In the case of making ammonia, the right hand side of the equation has the fewer molecules (2 molecules on the right versus 4 (3 + 1) molecules on the left). The forward reaction is exothermic.

The following table shows how the equilibrium position will shift with changes of pressure and temperature:

| The change | How the equilibrium shifts | Effect on the amount of ammonia |
|---|---|---|
| increase the pressure | equilibrium shifts in the direction that produces the smaller number of molecules of gas | equilibrium shifts to the right; more ammonia is produced |
| decrease the pressure | equilibrium shifts in the direction that produces the larger number of molecules of gas | equilibrium shifts to the left; less ammonia is produced |
| increase the temperature | equilibrium moves in the endothermic direction | equilibrium moves to the left; less ammonia produced |
| decrease the temperature | equilibrium moves in the exothermic direction | equilibrium shifts to the right; more ammonia is produced |

*equilibrium shifts to the right when the pressure is increased and the temperature is decreased*

$$3H_2(g) + N_2(g) \rightleftharpoons 2NH_3(g)$$

*equilibrium shifts to the left when the pressure is decreased and the temperature is increased*

## Examination Questions

**1** A student made some copper(II) sulfate solution by adding copper(II) carbonate to dilute sulfuric acid.

***a)*** (i) Complete the equation by writing the correct state symbol after each formula.

$CuCO_3(\_\_\_\_) + H_2SO_4(\_\_\_\_) \rightarrow CuSO_4(\_\_\_\_) + H_2O(\_\_\_\_) + CO_2(\_\_\_\_)$ *(2)*

**(ii)** State the colours of the copper compounds in the equation, that is, copper(II) carbonate and copper(II) sulfate. *(2)*

**(iii)** Apart from a colour change, what does the student **see** during the reaction? *(1)*

***b)*** Use words from the box to complete the sentences. The sentences explain the method the student used to make copper(II) sulfate.

Each word may be used once, more than once or not at all.

| | | | |
|---|---|---|---|
| **an acid** | **an alkali** | **a carbonate** | |
| **neutralisation** | **oxidation** | **reduction** | **a salt** |

When ____________ reacts with ____________ the solution formed contains ____________. The type of reaction occurring is ____________. *(4)*

***c)*** The teacher told the student to add an excess of copper(II) carbonate and remove it after the reaction had finished.

**(i)** Why is an excess of copper(II) carbonate added? *(1)*

**(ii)** How is the excess copper(II) carbonate removed after the reaction has finished? *(1)*

***d)*** What method would the student use to obtain a sample of copper(II) sulfate crystals, $CuSO_4.5H_2O(s)$, from the solution formed in the reaction? *(1)*

***(Total 12 marks)***

**2** A teacher described a reaction as follows:

"When zinc is added to dilute sulfuric acid, hydrogen gas and a zinc compound are formed."

***a)*** Write a **word** equation for this reaction. *(1)*

***b)*** A student added a piece of zinc to a test tube containing dilute sulfuric acid. Use words from the box to complete an account of the reaction.

Each word may be used once, more than once or not at all.

| | | |
|---|---|---|
| **effervescence** | **endothermic** | **exothermic** |
| **faster** | **precipitation** | **slower** |

**(i)** After adding the zinc to the acid the student saw ____________.

**(ii)** After a few minutes the student noticed that the reaction was ____________ than at the start.

**(iii)** The test tube was warmer at the end of the reaction than at the start. This showed that the reaction was ____________. *(3)*

***c)*** Describe a test the student could use to show that the gas formed was hydrogen. *(2)*

***d)*** At the end of the reaction there was some solid zinc left in the test tube. The student removed the zinc, leaving a colourless solution of zinc sulfate.

(i) Which technique did the student use to remove the zinc at the end of the reaction? *(1)*

(ii) The student asked the teacher how to test the colourless solution to find out if it contained sulfate ions. The teacher wrote this equation:

$ZnSO_4(aq) + BaCl_2(aq) \rightarrow BaSO_4(s) + ZnCl_2(aq)$

State the name of the compound added to the colourless solution and describe what the student would see. *(2)*

***(Total 9 marks)***

**3** Hydrogen peroxide decomposes into water and oxygen.

$2H_2O_2 \rightarrow 2H_2O + O_2$

The reaction is very slow but becomes faster if manganese(IV) oxide is added. The manganese(IV) oxide does not get used up during the reaction.

***a)*** What is the role of the manganese(IV) oxide in this reaction? *(1)*

***b)*** The graph shows how the volume of oxygen collected changed with time when 1 g of small lumps of manganese(IV) oxide were added to 10 cm³ of hydrogen peroxide.

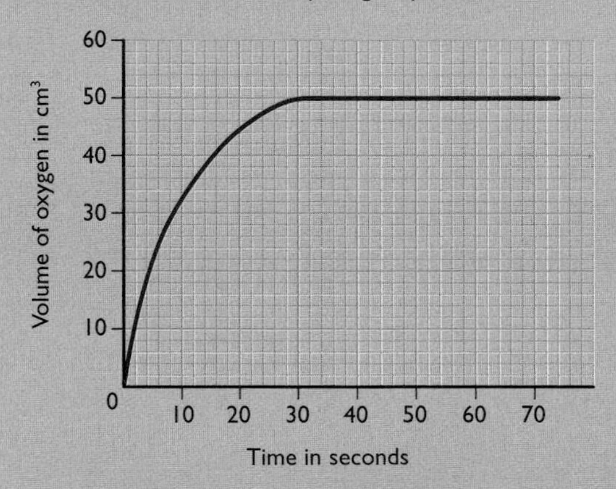


Sketch on the axes above the results obtained when:

(i) the experiment is repeated using 1 g of powdered manganese(IV) oxide. Label this sketch **A**. *(2)*

(ii) the same volume of hydrogen peroxide is used but 5 cm³ of water is added to it before the manganese(IV) oxide is added. Label this sketch **B**. *(2)*

***c)*** Describe a test for oxygen gas. *(2)*

***(Total 7 marks)***

**4** A common example of an exothermic reaction is the complete combustion of methane, as shown in the equation.

$CH_4(g) + 2O_2(g) \rightarrow CO_2(g) + 2H_2O(g)$

This reaction can be represented by an energy level diagram. Complete the diagram by showing the products of the reaction. *(1)*

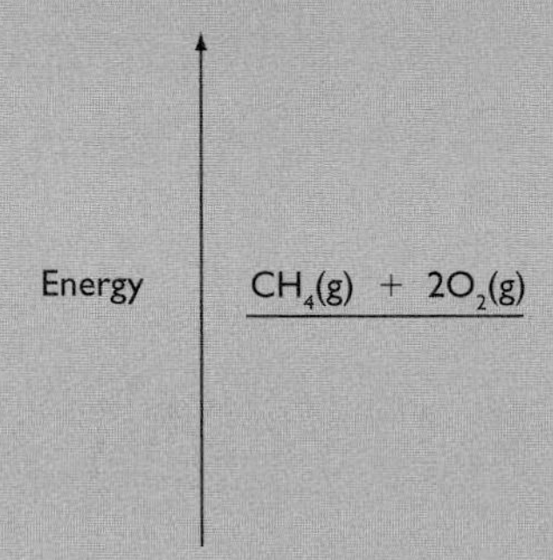


***b)*** The table shows the average values of some bond dissociation energies.

| Bond | C–H | O–H | O=O | C=O |
|---|---|---|---|---|
| Dissociation energy (kJ / mol) | 412 | 463 | 496 | 743 |

Methane and water contain only single bonds. Oxygen and carbon dioxide contain only double bonds.

Use the values in the table to calculate the energy change occurring during the complete combustion of methane. *(3)*

***c)*** At room temperature the reaction between methane and oxygen is very slow. State **three** different changes in conditions that would increase the rate of this reaction. *(3)*

***d)*** Another reaction of methane, used in industry, is shown by the equation:

$CH_4(g) + H_2O(g) \rightleftharpoons CO(g) + 3H_2(g)$
$\Delta H = +210\,kJ/mol$

(i) What do the symbols $\rightleftharpoons$ and $\Delta H$ represent? *(2)*

(ii) The reaction is carried out at 2 atmospheres pressure and 1000°C. Predict what would happen to the amounts of carbon monoxide and hydrogen formed if the pressure was increased at constant temperature and if the temperature was decreased at constant pressure. *(2)*

***(Total 11 marks)***

Notes

# Chapter 25: Extraction and uses of metals

A rock that contains enough metal to make it worth mining is called an **ore**.

Only the most unreactive metals, such as silver and gold, occur as **elements** in their ores.

The rest of the metals are found as **compounds**, from which the metals have to be extracted.

## Methods of extraction

The method of extraction is linked to the position of the metal in the reactivity series, since the more reactive metals are better at 'keeping hold' of the elements in their compounds. This means that the more reactive the metal, the more difficult and expensive it is to extract.

The table shows the link between the position of a metal in the reactivity series and the method used to extract it from its compounds:

| Metal | Method of extraction |
|---|---|
| potassium<br>sodium<br>lithium<br>calcium<br>magnesium<br>aluminium | electrolysis of the molten chloride or molten oxide<br>(electrolysis is the most powerful method of extraction but is very expensive since it uses a lot of electricity) |
| carbon | |
| zinc<br>iron<br>copper | heat with a reducing agent such as carbon or carbon monoxide |
| silver<br>gold | occur naturally as the elements |

### Extraction of aluminium

You do not have to learn to draw this diagram, but you may be asked to label it.

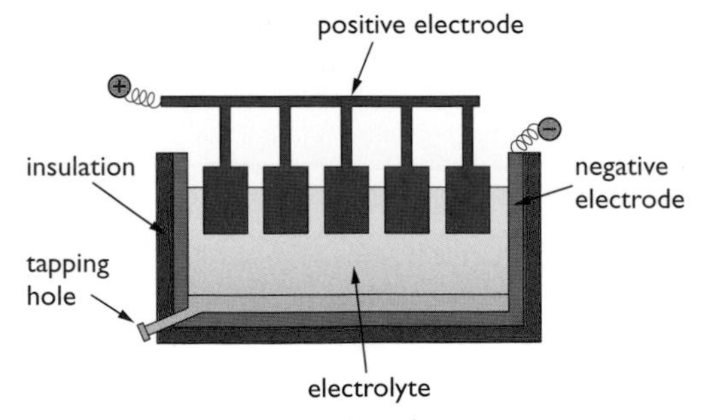


**Figure 25.1** *Extraction by electrolysis.*

- The positive electrode is made of graphite (carbon).
- The negative electrode is made of graphite (carbon).
- The electrolyte is a solution of aluminium oxide dissolved in molten cryolite.

### *Essential information:*

- The main ore of aluminium is bauxite.
- The bauxite is first purified to produce aluminium oxide, $Al_2O_3$.
- Aluminium oxide has a very high melting point and hence it is dissolved in molten cryolite to make the electrolyte. This mixture has a much lower melting point and is also better a conductor of electricity than molten aluminium oxide.
- The reaction at the negative electrode is:

$$Al^{3+} + 3e^- \rightarrow Al$$

- The aluminium melts and collects at the bottom of the cell. It is then tapped off.
- The reaction at the positive electrode is:

$$2O^{2-} \rightarrow O_2 + 4e^-$$

- Some of the oxygen produced at the positive electrode then reacts with the graphite (carbon) to produce carbon dioxide gas:

$$C(s) + O_2(g) \rightarrow CO_2(g)$$

This means that the positive electrode gradually burns away and needs to be replaced at regular intervals.

## Extraction of iron

### *Essential information:*

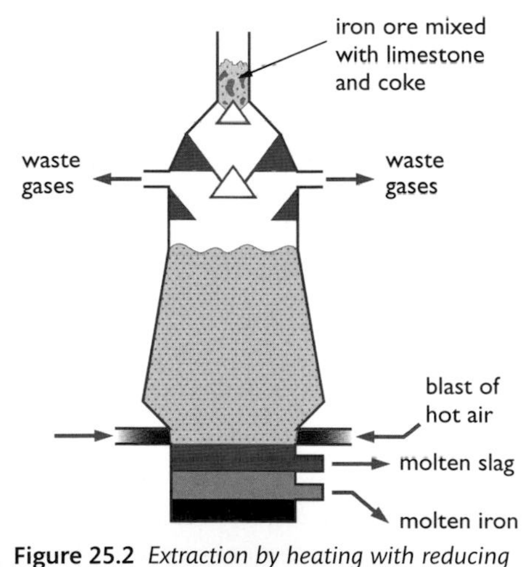


**Figure 25.2** *Extraction by heating with reducing agent.*

- The raw materials are iron ore (haematite), coke (carbon), limestone (calcium carbonate) and air.
- Iron ore, coke and limestone are mixed together and fed into the top of the blast furnace.
- Hot air is blasted into the bottom of the furnace.

You do not have to learn to draw this diagram, but you may be asked to label it.

- Oxygen in the air reacts with the coke to form carbon dioxide:

$$C(s) + O_2(g) \rightarrow CO_2(g)$$

- Carbon dioxide reacts with coke to form carbon monoxide:

$$CO_2(g) + C(s) \rightarrow 2CO(g)$$

- Carbon monoxide reduces the iron(III) oxide in the iron ore:

$$Fe_2O_3(s) + 3CO(g) \rightarrow 2Fe(l) + 3CO_2(g)$$

- The iron melts and collects at the bottom of the furnace, where it is tapped off.
- The calcium carbonate in the limestone decomposes to form calcium oxide:

$$CaCO_3(s) \rightarrow CaO(s) + CO_2(g)$$

It is important to remove the silicon dioxide since it would remain as a solid in the furnace and would eventually clog it up. This would mean that the furnace would have to be shut down for a time whilst the solid was removed. This would be very costly.

- The calcium oxide reacts with silicon dioxide, which is an impurity in the iron ore, to form calcium silicate:

$$CaO(s) + SiO_2(s) \rightarrow CaSiO_3(l)$$

- The calcium silicate melts and collects as a molten **slag** on top of the molten iron, which is then tapped off separately.

## The uses of aluminium and iron

The tables list some of the major uses of aluminium and iron together with the most important property related to the use given.

### *Aluminium*

| Use | Most important property |
|---|---|
| aeroplane bodies | high strength-to-weight ratio |
| overhead power cables | good conductor of electricity |
| saucepans | good conductor of heat |
| food cans | non-toxic |
| window frames | resists corrosion |

### *Iron*

| Use | Most important property |
|---|---|
| car bodies | strong (withstands collisions) |
| iron nails | strong |
| ships, girders and bridges | strong |

Although the uses are listed for aluminium and iron, the metals are rarely used in pure form. Pure aluminium is not very strong, so aluminium alloys are normally used instead. The aluminium can be strengthened by adding other elements such as silicon, copper or magnesium.

Similarly, most iron is used in the form of mild steel, which is an alloy containing a little carbon.

Pure iron, often called **wrought** iron, is soft compared with mild steel and is also very malleable. It is not strong enough to use to make car bodies, etc. Its major use is in making ornamental work for gates and fences.

**EXAMINER'S TIP**

- It is very important when asked to give a property of a metal that relates to its use, you pick the **most important** property. For example, aluminium is ductile (can be drawn into a wire), which is necessary to make power cables, but it is **not** the most important property.
- Be careful in your use of words when describing a property. For example, you will be allowed to give **low density** as an alternative to high strength-to-weight ratio for making aeroplanes, but stating that aluminium is **light** will be marked wrong. Another example is 'aluminium does not corrode'. This is often stated in textbooks as a property of aluminium. It is not true. Aluminium does corrode but very slowly, so it is better to say that aluminium 'resists corrosion'.

# Chapter 26: Crude oil

## What is crude oil?

Crude oil is a thick, sticky, black liquid that is found under the ground and under the sea in certain parts of the world such as the Middle East and Texas, in the USA.

It is a mixture of **hydrocarbons**, mostly alkanes ***(see Chapter 18)***.

## Refining of crude oil

Crude oil, as such, has no direct use. It has to be **refined** before it is of any use. The first step in the refining of crude oil is **fractional distillation**.

Fractional distillation is carried out in a **fractionating column.** The column is hot at the bottom and gradually becomes cooler towards the top.

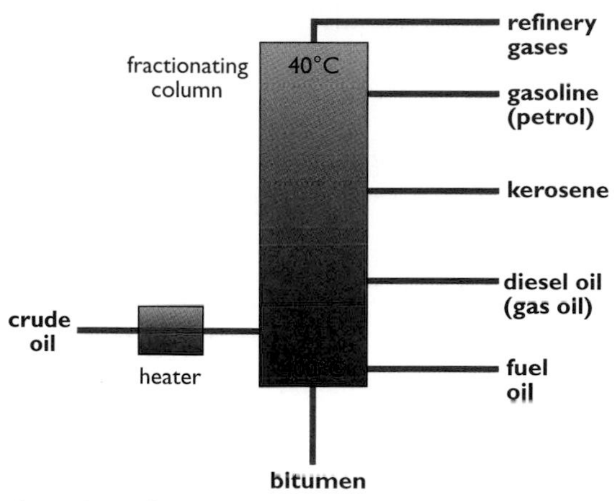


**Figure 26.1** *The industrial separation of crude oil.*

The crude oil is split into various **fractions** as described below. A fraction is a **mixture** of hydrocarbons with very **similar boiling points.**

- Crude oil is heated to convert it into a vapour. The vapour is then fed into the bottom of the fractionating column.
- The hydrocarbons with very high boiling points (fuel oil and bitumen) immediately turn into liquids and are tapped off at the bottom of the column.
- The hydrocarbons that have boiling points lower than 400°C remain as gases and rise up the column. As they rise they cool down.
- The different fractions will condense at different heights according to their different boiling points. When they condense they are tapped off as liquids.
- The fraction with the lowest boiling point (refinery gases) remains as a gas and comes out at the top of the column.

**EXAMINER'S TIP**

Different sources will give different fractions for the distillation of crude oil. This is for a variety of reasons, one of which is that the exact composition of crude oil depends on where in the world it comes from. The fractions given in the table below are the same as those given in the Edexcel International GCSE specification. It is best for you to learn these as they will always be marked correct in your answers to exam questions.

### Properties of the different fractions

The table lists some of the properties of the different fractions.

| Fraction | Change in number or carbon atoms in molecule | Change in boiling point | Change in viscosity |
|---|---|---|---|
| refinery gases<br>gasoline<br>kerosene<br>diesel oil<br>fuel oil<br>bitumen | ↓ increases | ↓ increases | ↓ increases |

The larger the number of carbon atoms in a molecule, the longer the length of the carbon chain. For this reason the higher boiling point fractions are said to the **long-chain** hydrocarbons, whilst the lower boiling point fractions contain the **short-chain** hydrocarbons.

## Uses of the fractions

The table lists some of the major uses of each fraction:

| Fraction | Use |
|---|---|
| refinery gases | bottled gas for camping, etc. |
| gasoline | petrol for cars |
| kerosene | fuel for aeroplanes; 'oil' for central heating boilers in the home; 'paraffin' for small heaters and lamps |
| diesel oil | diesel fuel for buses, lorries, trains and cars |
| fuel oil | fuel for ships and for industrial heating |
| bitumen | road surfaces and covering flat roofs of buildings |

Although there is a use for each fraction, the amount of the higher boiling point fractions in crude oil is far greater than is needed. The amount of gasoline fraction in crude oil is far less than is needed. For this reason, chemists devised a method of converting the long-chain hydrocarbons in the higher boiling point fractions into shorter-chain hydrocarbons, in order to make more gasoline, and so make more petrol for cars. This process is called **cracking**.

## How does cracking work?

The long-chain hydrocarbon (alkane) molecules are passed over a catalyst (either silica or aluminium oxide) heated to about 600–700°C. The long-chain hydrocarbons break down into short-chain alkane molecules and at least one alkene molecule.

A typical example is the cracking of decane ($C_{10}H_{22}$) to produce octane ($C_8H_{18}$) and ethene ($C_2H_4$):

$$C_{10}H_{22}(g) \rightarrow C_8H_{18}(g) + C_2H_4(g)$$

The octane produced can be used to make petrol. The ethene can be used to make poly(ethene), a polymer ***(see Chapter 27)***.

There are, though, many different products that can be obtained by the cracking of decane. Another possibility is:

$$C_{10}H_{22}(g) \rightarrow C_7H_{16}(g) + C_3H_6(g)$$

It can even produce hydrogen:

$$C_{10}H_{22}(g) \rightarrow C_7H_{14}(g) + C_3H_6(g) + H_2(g)$$

Remember also that the fraction which is being cracked contains not just decane, but a number of other alkane molecules, since it is a mixture. This means that the final product will contain a lot of compounds that will need careful and expensive separating into pure compounds before they can be used.

## Burning fuel in cars

In Chapter 18 it was mentioned that the incomplete combustion of an alkane produces carbon monoxide. The same is true when petrol or diesel is burnt in cars; some of the fuel is not completely burnt because of a lack of oxygen.

The carbon monoxide produced passes out through the exhaust pipe of the car and gets into the atmosphere. This is potentially dangerous, since carbon monoxide is poisonous to humans as it reduces the capacity of the blood to carry oxygen.

The temperature in a car engine is high enough for nitrogen and oxygen from the air to react to form oxides of nitrogen. These oxides are also passed out through the exhaust of the car and when they get into the atmosphere they can dissolve in the water in the air to form acid rain.

A lot of cars are now fitted with **catalytic convertors**, which are placed in the exhaust system of the car. These convertors attempt to convert as much carbon monoxide as possible into carbon dioxide, and also reconvert oxides of nitrogen into nitrogen and oxygen.

# Chapter 27: Synthetic polymers

## Addition polymers

An addition polymer is a long-chain molecule that has been formed when many small molecules (monomers) have been joined together. The process of making an addition polymer is called **addition polymerisation**. The monomer molecule contains a carbon–carbon double bond. During addition polymerisation, one of the double bonds in the monomer breaks and this allows the monomer molecules to join together to form a long chain of carbon atoms all bonded together.

The simplest monomer is ethene and this forms poly(ethene) when it polymerises. An equation that represents the polymerisation of ethene is:

$$n\ CH_2{=}CH_2 \rightarrow -(-CH_2-CH_2-)_n-$$

Using displayed formulae, the equation becomes:

$$n\ H_2C{=}CH_2 \longrightarrow \left[\begin{array}{c} H\ \ H \\ | \ \ \ | \\ -C-C- \\ | \ \ \ | \\ H\ \ H \end{array}\right]_n$$

The letter 'n' represents a very large, but variable, number. It simply means that the structure in the brackets repeats itself many times in the molecule. For this reason, the structure in the brackets is called the **repeat unit** of poly(ethene).

**EXAMINER'S TIP**

You may be asked how you know that poly(ethene) is an addition polymer. The answer is that when it is made, it is the only product of the reaction. (Compare this with condensation polymerisation.)

The table gives the formulae of some monomers together with their repeat unit:

| Name of monomer | Formula of monomer | Name of polymer | Formula of repeat unit |
|---|---|---|---|
| propene | $CH_3CH{=}CH_2$ (H₃C–CH=CH₂ displayed: C bonded to H, H, H; C=C; H; H, H) | poly(propene) | $-C(CH_3)(H)-C(H)(H)-$ |
| chloroethene (vinyl chloride) | $H_2C{=}CHCl$ | poly(chloroethene) (polyvinyl chloride, or PVC for short) | $-C(H)(H)-C(Cl)(H)-$ |
| tetrafluoroethene | $F_2C{=}CF_2$ | poly(tetrafluoroethene) | $-C(F)(F)-C(F)(F)-$ |

**EXAMINER'S TIP**

You need to know how to write a repeat unit using the formula of any monomer given to you, and also how to convert a repeat unit back into its monomer. You are therefore advised to practise this with as many different monomers as you can find and then check your answers with your teacher.

## Uses of some addition polymers

1. Poly(ethene)

Poly(ethene) is resistant to chemical attack and can therefore be used to store food, drinks and other chemicals, including acids and alkalis. There are a number of forms of poly(ethene) including high density (HDPE), medium density (MDPE) and low density poly(ethene) (LDPE). Each form has its own range of uses. Some of these are listed below.

HDPE – milk jugs, detergent bottles, margarine tubs, garden furniture

MDPE – gas pipes, rubbish ('wheelie') bins, storage tanks for fuel

LDPE – plastic bags, cling film

2. Poly(propene)

Poly(propene) is stronger and more hard wearing than poly(ethene). Its major uses include making food packaging, ropes and carpets.

3. Poly(chlorethene)

Poly(chlorethene) is tougher than poly(ethene), very hard wearing and more stable to heat. Its major uses include plastic sheets, artificial leather (for clothes and handbags etc), drainpipes and gutters, insulation for electrical wires and casings for electrical plugs.

## Disposing of addition polymers

One of the useful properties of addition polymers is that they are unreactive, so they are suitable for storing food and chemicals safely. Unfortunately, this property makes it difficult to dispose of these polymers. They are often buried in landfill sites or incinerated (burned).

### *Landfill*

Waste polymers are disposed of in landfill sites. This uses up valuable land as addition polymers tend to be non-biodegradable. This means microorganisms cannot break them down. They may last for many years in rubbish dumps, so the sites quickly fill up and new sites have to be found.

### *Incineration*

Polymers release a lot of heat energy when they burn. This energy can be used to heat homes or generate electricity. But there are problems with incineration. Carbon dioxide is produced, which may be contributing to climate change. Toxic gases can also be produced.

### *Recycling*

Many polymers can be recycled. This reduces disposal problems and the amount of crude oil used. But first the different polymers must be separated from each other. This can be difficult and expensive.

## Condensation reactions

A condensation reaction is one in which two molecules react together to form a new, larger molecule with the elimination of a small molecule such as water.

An example of a condensation reaction is the reaction between an alcohol, such as ethanol, and a carboxylic acid (an organic compound containing a –COOH group). This reaction is also known as esterification, since the compound produced is an ester.

$$\underset{\text{acid}}{-\overset{\overset{\displaystyle O}{||}}{C}-OH} \quad \underset{\text{alcohol}}{H-O-} \longrightarrow \underset{\text{ester}}{-\overset{\overset{\displaystyle O}{||}}{C}-O-} + H_2O$$

water eliminated

Another example of a condensation reaction is one that takes place between an amine (an organic compound containing an $-NH_2$ group) and a carboxylic acid. This time the compound formed is called an amide.

$$\underset{\text{acid}}{-\overset{\overset{\displaystyle O}{||}}{C}-OH} \quad \underset{\text{amine}}{H-\overset{\overset{\displaystyle H}{|}}{N}-} \longrightarrow \underset{\text{amide}}{-\overset{\overset{\displaystyle O}{||}}{C}-\overset{\overset{\displaystyle H}{|}}{N}-} + H_2O$$

water eliminated

### Condensation polymers

A condensation polymer is a polymer that is formed via a condensation reaction. In these reactions many monomers join together to make the polymer (a long-chained molecule) and a small molecule such as water is eliminated for each pair of monomers that join together. There are two main types of condensation polymers: polyesters and polyamides. A typical polyester is **terylene**. A typical polyamide is **nylon**.

# **Chapter 28:** The industrial manufacture of chemicals

## The manufacture of ammonia – the Haber process

Ammonia is made by combining nitrogen and hydrogen. The hydrogen is obtained either from natural gas or the cracking of hydrocarbons ***(see Chapter 26)***. The nitrogen is obtained from air.

$$3H_2(g) + N_2(g) \rightleftharpoons 2NH_3(g) \quad \Delta H = -92\ \text{kJ/mol}$$

The reaction is reversible and can reach a position of equilibrium ***(see Chapter 24)***. The equilibrium position can be shifted to the right hand side (to give a high yield of ammonia) by using a **low** temperature and a **high** pressure. However, the reaction is very slow at low temperatures, so a compromise of about 450°C is used. This is a compromise between rate and yield.

Carrying out the reaction at **very** high pressures is expensive since more energy would be required to compress the gases. Also, it is expensive to build the thick-walled pipes and reaction containers that would be necessary to withstand these very high pressures. Therefore a pressure of 200 atmospheres is used, which is still high but is not too expensive.

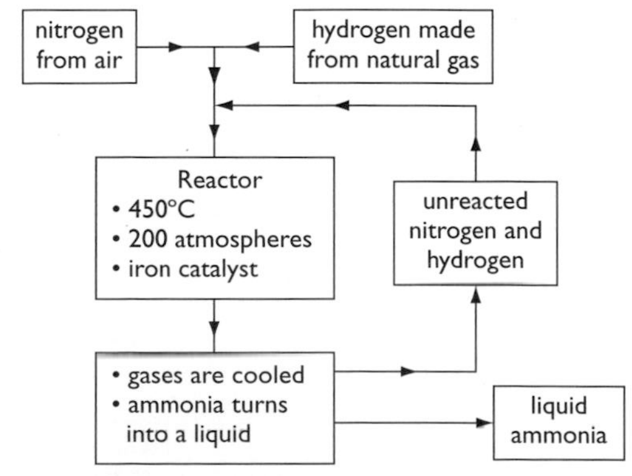


**Figure 28.1** *Flowchart of the Haber process.*

The reaction is still fairly slow at 450°C unless a catalyst is used. There are a number of metals that will catalyse this reaction, but the most economical one is iron. Under these conditions, about 15% of the hydrogen and nitrogen is converted into ammonia. The mixture leaves the reaction vessel and is then cooled. The ammonia liquefies and is tapped off. The hydrogen and nitrogen remain as gases and are mixed with more hydrogen and nitrogen and passed again into the reaction chamber.

### Uses of ammonia

- Manufacture of fertilisers.
- Manufacture of nitric acid.
- Manufacture of nylon.

## The manufacture of sulfuric acid – the contact process

Sulfuric acid is manufactured in the contact process, which has several stages:

**Stage 1:** Making sulfur dioxide, $SO_2$. Most sulfur dioxide is made by burning sulfur in air:

$$S + O_2 \rightarrow SO_2$$

**Stage 2:** Making sulfur trioxide, $SO_3$. To do this, sulfur dioxide is reacted with oxygen:

$$2SO_2(g) + O_2(g) \rightleftharpoons 2SO_3(g) \qquad \Delta H = -196 \text{ kJ/mol}$$

**Stage 3:** Making sulfuric acid, $H_2SO_4$. Sulfur trioxide is absorbed into concentrated sulfuric acid to make oleum, $H_2S_2O_7$:

$$H_2SO_4(l) + SO_3(g) \rightarrow H_2S_2O_7(l)$$

Oleum is then carefully diluted to make a concentrated solution of sulfuric acid:

$$H_2S_2O_7(l) + H_2O(l) \rightarrow 2H_2SO_4(aq)$$

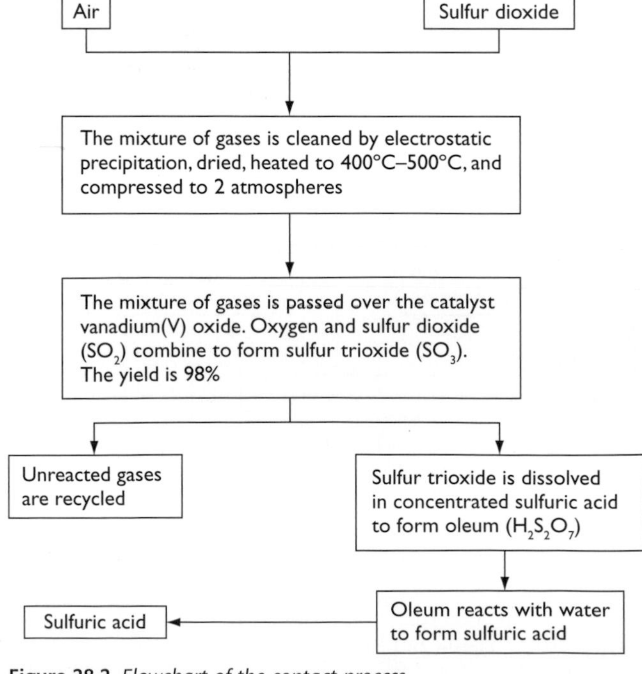


**Figure 28.2** *Flowchart of the contact process*

The reaction in stage 2 of the process is the one that gives the process its name, since the two gases react together **in contact** with a solid catalyst, vanadium(V) oxide. It is also a reversible reaction involving gases. The position of equilibrium, and hence the yield of sulfur trioxide, can therefore be affected by temperature and pressure.

Since the forward reaction is exothermic, the yield of sulfur trioxide would be increased by using a low temperature. However, the reaction would be slow at low temperatures, so the reaction is carried out at a temperature of about 450°C.

There are fewer molecules of gas on the right hand side (2) than on the left (3), so the yield of sulfur trioxide would be increased by using a high pressure. However, the yield is already very high (about 98%) at a pressure of 2 atmospheres, so using higher pressures would not be economical.

## Uses of sulfuric acid

- Manufacture of detergents.
- Manufacture of fertilisers.
- Manufacture of paints.

# The manufacture of sodium hydroxide and chlorine

Sodium hydroxide and chlorine are manufactured together in a single process, by the electrolysis of a concentrated solution of sodium chloride (brine). Hydrogen is a useful by-product of the electrolysis.

At the positive electrode (anode), chloride ions lose electrons to form chlorine molecules:

$$2Cl^-(aq) \rightarrow Cl_2(g) + 2e^-$$

At the negative electrode (cathode), water molecules gain electrons to form hydroxide ions and hydrogen molecules:

$$2H_2O(l) + 2e^- \rightarrow 2OH^-(aq) + H_2(g)$$

chlorine
hydrogen
concentrated sodium chloride solution
titanium anode
diaphragm
steel cathode
sodium hydroxide solution contaminated with sodium chloride

**Figure 28.3** *Apparatus for the manufacture of sodium hydroxide and chlorine.*

The overall equation for the reactions taking place is:

$$2NaCl(aq) + 2H_2O(l) \rightarrow 2NaOH(aq) + H_2(g) + Cl_2(g)$$

The diaphragm (see diagram above right) is necessary to keep apart sodium hydroxide and chlorine. If they mix they will react together to make bleach (see uses of sodium hydroxide and chlorine below).

## Uses of sodium hydroxide and chlorine

| Sodium hydroxide | Chlorine |
|---|---|
| manufacture of bleach | sterilising water supplies |
| manufacture of paper | manufacture of bleach |
| manufacture of soap | manufacture of hydrochloric acid |
| manufacture of detergents | manufacture of PVC |

**EXAMINER'S TIP**

It is important to state that sodium hydroxide and chlorine are used to **manufacture** the materials mentioned. For example, sodium hydroxide is not used as a bleach; it is used to **make** bleach. The other common mistake is to say that chlorine is used to **purify** water. Adding chlorine to water makes it **less pure**, but it does, however, kill the bacteria and other harmful micro-organisms in the water.

## Examination Questions

**1** *a)* Ammonia is made industrially by the Haber process. In this process nitrogen is reacted with hydrogen. The flow diagram shows what happens.

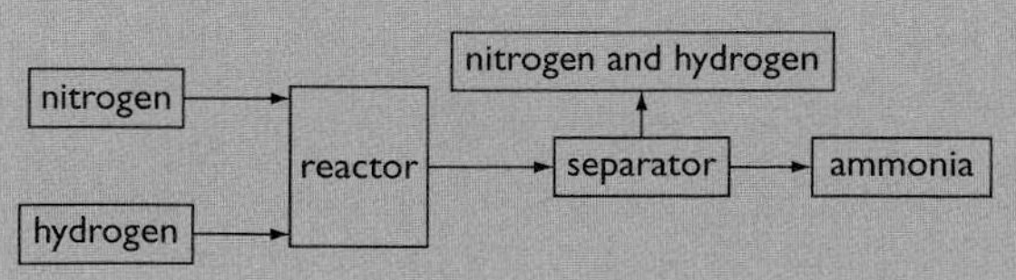


(i) Give the names of the raw materials from which the nitrogen and hydrogen are obtained. *(2)*

(ii) State the conditions used in the reactor. *(3)*

(iii) How is the ammonia separated from the unreacted nitrogen and hydrogen? *(1)*

(iv) What is done with the unreacted nitrogen and hydrogen? *(1)*

**b)** Ammonium nitrate can be used as a fertiliser to increase plant growth. It is made by reacting ammonia solution with nitric acid. Write a chemical equation for this reaction. *(2)*

***(Total 9 marks)***

**2** Crude oil is a mixture of hydrocarbons. The mixture can be separated into fractions by the process of fractional distillation.

**a)** Fractional distillation of crude oil produces the fractions bitumen, diesel, fuel oil, gasoline, kerosene and refinery gases. State **one** use of bitumen and **one** use of kerosene. *(2)*

**b)** Gasoline is used as a fuel for cars. When gasoline undergoes complete combustion the products are carbon dioxide and water.

**(i)** Write a word equation for the complete combustion of gasoline. *(1)*

**(ii)** In car engines, incomplete combustion takes place. Why is the combustion incomplete? *(1)*

**(iii)** Explain why the incomplete combustion of gasoline can be harmful to humans. *(3)*

**c)** Fractional distillation works because each fraction has a different boiling range. Describe how you could obtain a fraction with a boiling range of 80°C to 120°C **in the laboratory** from a sample of crude oil. Name the items of apparatus you would need. *(3)*

***(Total 10 marks)***

**3** Crude oil is a mixture of hydrocarbons.

**a)** Which **two** elements are present in the compounds in crude oil? *(2)*

**b)** Crude oil is separated into fractions by heating and passing the vapour into a fractionating column. Explain why the fractions separate in the column. *(2)*

**c)** Two of the fractions are gasoline and bitumen. Give **one** use of each. *(2)*

**d)** Name **two** fractions formed in the fractional distillation of crude oil, other than gasoline and bitumen. *(2)*

**e)** (i) Identify the **two** products of **complete** combustion of hydrocarbons. *(2)*

**(ii)** Explain why the **incomplete** combustion of hydrocarbons is harmful to humans. *(2)*

***(Total 12 marks)***

**4 a)** Iron is extracted from iron oxide in a blast furnace.

**(i)** Name the two solid raw materials added to the top of the blast furnace with the iron oxide. *(2)*

**(ii)** Molten iron collects at the bottom of the furnace. Which molten substance collects above molten iron? *(1)*

**(iii)** The word equation for one reaction that occurs in the blast furnace is:

carbon + oxygen → carbon dioxide

Write the chemical equation for this reaction, including state symbols. *(2)*

**(iv)** The chemical equation for another reaction that occurs in the blast furnace is:

$C + CO_2 \rightarrow 2CO$

Which substance in this equation is reduced? *(1)*

**b)** After some time, rust forms on many objects made from iron.

**(i)** Name **two** substances needed for rust to form on iron. *(2)*

**(ii)** Iron buckets can be prevented from rusting by galvanising. In this process the iron bucket is coated with another metal. Name the metal used to galvanise iron. Describe how this metal prevents the iron from rusting. *(2)*

**(iii)** Suggest why the iron inside a motor car engine does not rust. *(1)*

***(Total 11 marks)***

**5** The diagram shows a blast furnace used to extract iron from its ore. The name of one of the raw materials is shown.

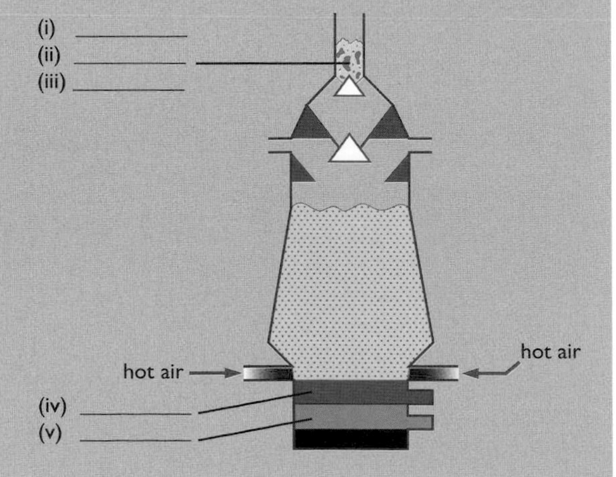

*a)* Complete the labelling of the diagram using the names or formulae of the substances. *(5)*

*b)* The word equations for two reactions occurring in the blast furnace are:

**Reaction 1**
carbon + oxygen → carbon dioxide

**Reaction 2**
carbon dioxide + carbon → carbon monoxide

(i) Which of these reactions (**1** or **2**) produces a high temperature in the blast furnace? *(1)*

(ii) State, with a reason, which substance in **reaction 2** undergoes reduction. *(2)*

*c)* Why is it important that carbon monoxide is **not** released into the atmosphere? *(1)*

*d)* Why is aluminium not extracted from its ore using a blast furnace? *(1)*

*(Total 10 marks)*

**6** The diagram shows how aluminium is extracted on an industrial scale.

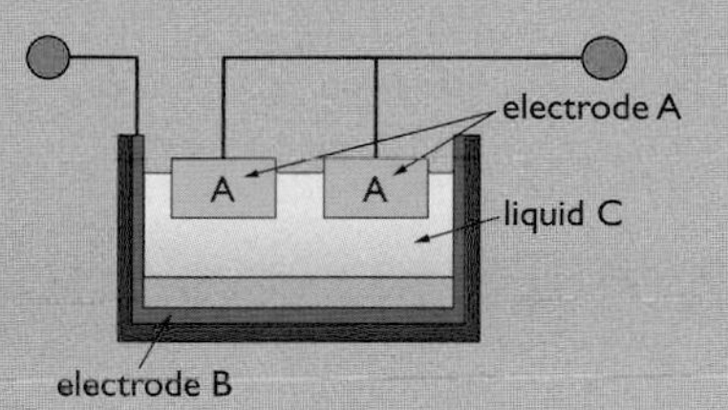


*a)* (I) Name the process used to extract aluminium. *(1)*

(ii) Name the material used for both electrodes **A** and **B**. *(1)*

(iii) Using the symbols + and − identify the polarity of the electrodes **A** and **B**. Write **one** symbol in each circle in the diagram above. *(1)*

(iv) Identify the **two** compounds present in liquid **C**. *(2)*

(v) State **one** major cost that makes this process more expensive than the extraction of iron. *(1)*

*b)* The mixture of gases coming from electrodes **A** contains an element and a compound.

(i) Identify the element. *(1)*

(ii) Identify the compound and explain how it forms. *(2)*

*(Total 9 marks)*

**7** This question is about the synthetic polymer nylon.

*a)* Poly(ethene) is an addition polymer. What type of polymer is nylon? *(1)*

*b)* Nylon can be made using the monomers **A** and **B** represented in the diagrams.

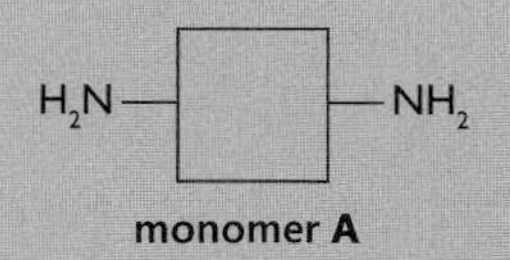


**monomer A**

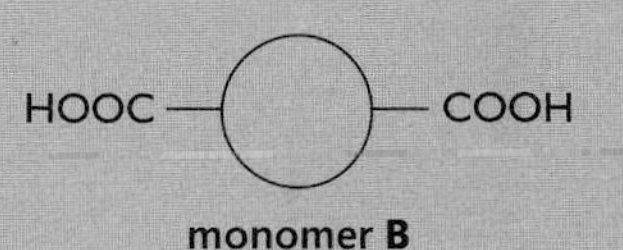


**monomer B**

(i) What type of compound is monomer **A**? *(1)*

(ii) What type of compound is monomer **B**? *(1)*

(iii) Draw a diagram to show the structure of the polymer formed from **A** and **B**. You must draw enough of the structure to make the repeat unit clear. *(3)*

*c)* Nylon has a simple molecular structure. Use words from the box to complete the sentences. Each word may be used once, more than once or not at all.

| | | |
|---|---|---|
| **ions** | **high** | **low** |
| **molecules** | **strong** | **weak** |

Nylon has a __________ melting point. This is because there are __________ forces between the __________ that make up the structure. *(3)*

*(Total 9 marks)*

## Notes

..........

..........

..........

..........

# Types of skills examined

There are three types of skill assessed in both Paper 1 and Paper 2. These are:

**AO1** – Knowledge and understanding

**AO2** – Application of knowledge and understanding, analysis and evaluation

**AO3** – Investigative skills

This section deals with AO3. Some International GCSE examination-style questions are presented with different answers and a discussion of these answers. One way to make use of these questions and answers is first of all to try to answer them yourself; then compare your answers with those given and finally read the commentaries.

The questions assessing practical skills may be part of an overall question that could also be assessing AO1 and/or AO2 skills, or may be contained within a single question designed specifically to assess AO3. In total, 20% of the marks allocated to each of Paper 1 and Paper 2 will be for the assessment of AO3.

In these questions you are expected to be able to:

- recognise and explain the use of common items of laboratory apparatus
- plan practical procedures
- use the idea of a fair test
- read scales to an appropriate degree of accuracy and perform simple mathematical operations (including finding the mean) on results obtained
- use correct units for values
- record results in tables and use data presented in a variety of formats
- draw and use bar charts
- plot and use graphs and draw straight lines of best fit and curves of best fit
- identify relationships from graphs
- comment on the repeatability and accuracy of data
- draw conclusions and offer explanations
- identify anomalous results and explain how they may have arisen
- evaluate provided procedures and suggest improvements
- suggest further experimental work that may be required.

# Explanation of terms

The questions assessing AO3 will contain terms that have a specific meaning to the examiner. These terms are explained in the following table:

You are not expected to be able to quote these meanings in an examination.

| Term | Meaning |
|---|---|
| true value | The value that would be obtained under ideal conditions. Values from data booklets can be considered to be true values. |
| accuracy | An accurate measurement is one that is close to the true value. |
| precision | If a repeated measurement of a value gives the same, or very similar, results each time, the value is precise. |
| anomalous readings | Anomalous readings are readings that fall outside the normal, or expected, range of measurements. They will show on a graph as a point, or points, standing clearly away from a line of best fit. |
| concordant readings | Concordant readings are obtained when any measurement is repeated and all the readings are close, or identical. In titration work at International GCSE, concordant readings are those within $\pm 0.20\ cm^3$ of one another. |
| mean (average) | A mean value for a set of measurements is calculated by adding together all the measurements and then dividing the total by the number of measurements made ***(see p. 101)***. |
| repeatability | A measurement is repeatable when, under the same conditions, an individual student or group of students obtains the same or similar results. |
| reproducibility | A measurement is reproducible if, under similar, but not necessarily identical, conditions different students or different groups of students obtain the same or similar results. |
| variable | A quantity that is being measured or controlled in an experiment. It can have different values. One example is time. |
| continuous variable | A variable that can have any number of intermediate values between those that are being measured. For example, if in an experiment measurements of a variable are being taken every 20 seconds, then it is also possible to take readings in between those values of time. Hence time is a continuous variable. |
| categoric (discontinuous/ discrete) variable | A variable that cannot have intermediate values. For example, atomic number is a categoric variable since it can have only whole-number values – it is not possible to have a fractional number of protons in the nucleus of an atom. |
| independent variable | The variable that is changed to see its effect on the dependent variable. For example, in an experiment to investigate the effect on the rate of reaction of changes in concentration of a solution, the concentration of solution is the independent variable. |
| dependent variable | The variable that changes as a result of changes to the independent variable. In the above example, rate of reaction is the dependent variable. |
| control variable | A variable that can affect the outcome of an investigation. For example, when investigating the effect of concentration of solution on rate of reaction, temperature is a control variable, since changes in temperature will also have an effect on the rate of reaction. |
| fair test | An investigation or experiment in which only the values of the independent variables are changed. The values of all other control variables have been kept constant. |

A measurement can be precise but inaccurate ***(see p. 101)***.

The term 'reliability' has commonly been used here in the past. It should now be replaced by repeatability. However, if reliability is used in an examination question, you should interpret it as meaning repeatability.

| Term | Meaning |
|---|---|
| data | Numerical values of the independent and dependent variables that are recorded during an experiment or investigation. Data is usually recorded in tables to make comparison easy. |
| correlation | The relationship between the independent and dependent variables in a given experiment or investigation. |
| positive correlation | The value of the dependent variable **increases** as the value of the independent variable also **increases**. For example, the rate of a reaction increases as the temperature increases. In a graph, a straight line or curve that slopes **upwards** shows positive correlation. |
| negative correlation | The value of the dependent variable **decreases** as the value of the independent variable **increases**. For example, in the reaction between magnesium and dilute hydrochloric acid, the mass of magnesium remaining decreases as the time increases. In a graph, a straight line or curve that slopes **downwards** shows negative correlation. |
| directly proportional | If a graph of the dependent variable plotted against the independent variable produces a straight line **that passes through the origin**, then the dependent variable is directly proportional to the independent variable. |
| line of best fit | A straight line or curve that passes through, or close to, as many of the plotted points as possible, and ignores the anomalous results. |
| range (of a variable, or results) | The difference between the minimum and maximum values of the independent and dependent variables. It is important to have as wide a range as possible in order to have confidence in the conclusion you are making from the results of the experiment or investigation. |
| degree of confidence in conclusions | A **qualitative** judgement, expressing the extent to which a conclusion is justified by the quality of evidence provided. In making this judgement, you need to take into account, for example, how the experiment has been performed and the degree of accuracy of the measurements taken. Thus, in an experiment to determine the relationship between the rate of a reaction and the concentration of solution, you might have assumed that each reaction mixture was at room temperature, and that room temperature stayed constant throughout the time you were carrying out your experiments. This may be or may not be a valid assumption, but it does put some doubt on the degree of confidence you can have in your conclusion. |
| valid conclusion | A conclusion is valid if it is supported by valid data, obtained from a 'fair test' experiment and is based on sound reasoning. |

In the examination, you will be told whether to draw a **straight line** or a **curve** of best fit.

## More on accuracy and precision

Remember, accuracy is how close the measurement you make is to the true value. Any measurement you make in an experiment or investigation will have some degree of error involved. For example, a thermometer that is graduated in 2°C intervals has an accuracy of ±1°C (half of the scale division). Hence a reading of 20°C could be 19°C or 21°C (i.e. the level of accuracy for this reading is ±5%).

As well as accuracy, precision is also important. A precise instrument gives a consistent reading when it is used repeatedly for the same measurements.

For example:
A beaker is weighed three times on balance A:
The readings are: 74 g, 77 g, 71 g
So the range is = 71 g − 77 g = 6 g

It is then weighed three times on balance B:
The readings are: 76 g, 77 g, 75 g
So the range is = 75 g − 77 g = 2 g

Balance B has better precision. Its readings are grouped closer together.

However, even though an instrument gives precise readings, it may not be accurate. For example, if the true mass of the beaker is 73 g, then balance A has produced a more accurate value, since the mean of the three readings (see below) is 74. The mean of the three readings taken from balance B is 76. 74 is closer to the true value of 73 than is 76.

## Calculating a mean (average)

Calculating a mean value from a set of data is usually a simple affair. For example, you have the following five readings of the volume of gas given off during five different experiments:

21.0 $cm^3$, 20.8 $cm^3$, 21.1 $cm^3$, 20.9 $cm^3$ and 21.2 $cm^3$

The mean of these values is $\frac{21.0 + 20.8 + 21.1 + 20.9 + 21.2}{5} = 21.0\ cm^3$

However, you must take care when calculating the mean of a set of values when one, or more, of the values is anomalous. This is most likely to occur in an examination question involving titration results.

You are expected to know that concordant values of titration results are those that are within 0.20 $cm^3$ of one another. Therefore, if you are asked to calculate the mean of the following three titration results:

25.20 $cm^3$, 25.50 $cm^3$, 25.00 $cm^3$

you should treat the 25.50 $cm^3$ value as anomalous and not use it in the calculation of the mean. The mean is therefore:

$\frac{25.20 + 25.00}{2} = 25.10\ cm^3$

Also, be very careful to give your final answer to the number of decimal places that the examiner has asked for. In the above example, the answer to two decimal places is 25.10, but the answer to one decimal place is 25.1.

Lastly, be careful when rounding up or down to give an answer to the correct number of decimal places. For example, 25.464 is **25.46** to two decimal places, whilst 25.465 is **25.47** to two decimal places.

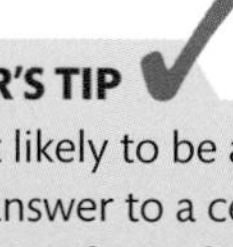

**EXAMINER'S TIP**

You are not likely to be asked to give an answer to a certain number of significant figures, but you will be penalised if you round up your final answer to too few significant figures, even if a specific number of decimal points has **not** been asked for. You need to look carefully at the data supplied to make a decision on the accuracy of your final answer.

## Making and recording measurements with appropriate precision

You are expected to be able to use a linear scale of an instrument to an accuracy of ± half of the smallest scale division and record the readings from the instrument to an appropriate level of precision. For example, look at the readings on the two thermometers shown below.

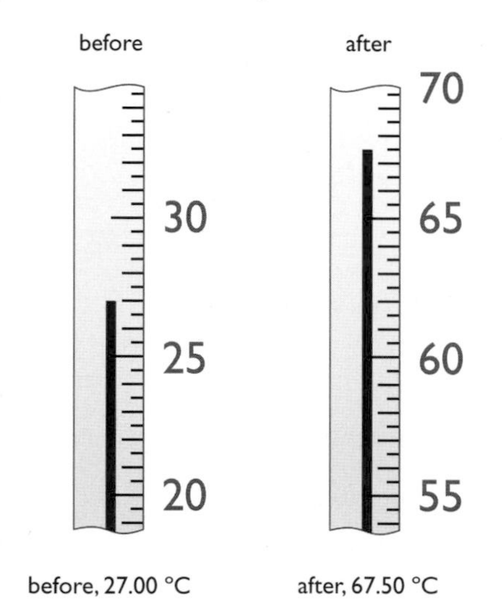


before, 27.00 °C    after, 67.50 °C

The second decimal place can be 0 or 5, since the scale can be read to a precision of ±0.25°C. A reading of, say, 67.75°C is possible to make. However, this level of precision of recording measurements is only asked for when using a burette. Therefore, if the readings on the thermometers above were given as 27.0°C and 67.5°C they would be marked as correct.

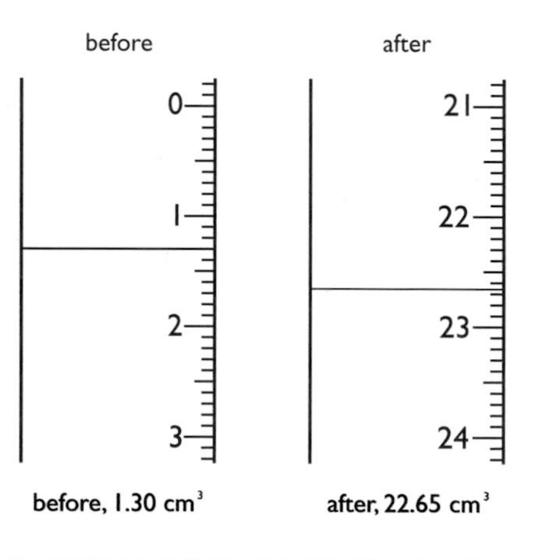


before, 1.30 $cm^3$    after, 22.65 $cm^3$

Now look at the example of burette readings (left). This time, 1.3 $cm^3$ for the 'before' reading would be marked as incorrect, and 22.6(0) $cm^3$ or 22.7(0) $cm^3$ would also be marked as incorrect for the 'after' reading.

The scale on a burette increases **downwards**. It is very easy to read the 'before' value here as 2.70 $cm^3$ and the 'after' value as 23.35 $cm^3$. Make sure you have this point very clear in your mind when taking readings from a burette.

The numerical data obtained from an experiment or an investigation should be presented in the form of a table, so that it can be analysed easily. Each column and row should have a heading together with the appropriate unit for the measurement. For example, the readings made during a titration could be recorded as:

| | practice titration | 1st accurate titration | 2nd accurate titration |
|---|---|---|---|
| Final burette reading in $cm^3$ | | | |
| Initial burette reading in $cm^3$ | | | |
| Volume of solution added in $cm^3$ | | | |

## Making and recording observations with appropriate precision

Observations are what you make during an experiment by using:

- your eyes (i.e. what you see)
- your ears (i.e. what you hear)
- your hands (i.e. what you feel).

Observations could also include what you can taste and smell, but this is not recommended in Chemistry since some substances are very harmful and even poisonous.

Most observations will be made with your eyes, but you may hear sounds such as hissing (such as when a piece of sodium is placed on water) or the noise caused by an explosion (such as the popping sound when hydrogen is mixed with air in a test tube and then ignited).

The common observations that you make with your eyes, and a description of what they mean, are given in the table below:

| Observation | Meaning |
|---|---|
| precipitate | An insoluble solid formed when a chemical reaction takes place in aqueous solution |
| bubbles/fizzing/ effervescence | The production of a gas when a chemical reaction is occurring in a liquid or a solution |
| solution | A clear liquid is produced. This observation would be most commonly recorded when a solid disappears after being added to a liquid or a solution |
| solid disappears | When only part or none of a solid remains visible after it has been added to a liquid or a solution |

It is important to state the colour of the precipitate formed.

Do not say 'gas given off' unless the gas is coloured, in which case you can say, for example, brown gas given off.

It is important to state the colour of the solution formed, only in most cases this will be **colourless**. It is not necessary say that the solution is clear, since **all** solutions are clear, even if they are coloured.

Here are some examples when the above observations can be made:

1. When an aqueous solution of silver nitrate is added to an aqueous solution of sodium chloride, a white precipitate is formed.
2. When magnesium ribbon is added to dilute hydrochloric acid in a test tube, the following observations are made:
   - bubbles are formed
   - the magnesium disappears
   - a colourless solution is formed
   - the test tube gets hot.

## Distinguishing between observations and deductions

It is important that, when you are asked to state the observations you would expect to make during a reaction, you do not list instead the deductions that can be made from the observations. For example, do not say a gas is evolved when the observation is bubbles. The exception to this is when the gas is coloured (see above table).

Other examples are in the table below:

| Observation | Deduction |
|---|---|
| Solid disappears | There are two possible deductions here:<br>1. the solid has dissolved, e.g. sodium chloride added to water, or<br>2. the solid has reacted to form a substance that dissolves in the liquid, e.g. when sodium is added to water, it reacts to form sodium hydroxide, which then dissolves in the water |
| Reaction mixture gets hot | The reaction or change is exothermic |
| Reaction mixture gets cold | The reaction or change is endothermic |

# Examination-style questions, answers and commentaries

## Worked Example 1

**Question**

Marble chips (calcium carbonate) react with hydrochloric acid.

The equation for the reaction is:

$CaCO_3(s) + 2HCl(aq) \rightarrow CaCl_2(aq) + H_2O(l) + CO_2(g)$

Some students investigated the rate at which carbon dioxide gas is given off at 25°C. In separate experiments they used different masses of the same-sized marble chips with the same volume of hydrochloric acid (an excess).

The students recorded these results.

Using 2.34 g of marble chips, 83 cm³ of carbon dioxide gas were collected in 60 seconds.

We got 45 cm³ of gas in 1 minute when we used 1.05 g of marble chips.

With 1.47 g of solid we made 98 cm³ of gas in 120 seconds.

In 60 seconds 0.59 g of solid gave 29 cm³ of carbon dioxide.

After 90 seconds, 1.21 g of calcium carbonate had made 54 cm³ of carbon dioxide.

Draw a suitable table and enter all of the results given and the units. *(3)*

**Answers**

*Student A:*

| Mass of marble chips (g) | Volume of gas collected (cm³) | Time taken (secs) |
|---|---|---|
| 2.34 | 83 | 60 |
| 1.05 | 45 | 1 |
| 1.47 | 98 | 120 |
| 0.59 | 29 | 60 |
| 1.21 | 54 | 90 |

*Student B:*

| Mass of marble chips | Volume of gas collected | Time taken |
|---|---|---|
| 2.34 g | 83 cm³ | 60 s |
| 1.05 g | 45 cm³ | 60 s |
| 1.47 g | 98 cm³ | 120 s |
| 0.59 g | 29 cm³ | 60 |
| 1.21 g | 54 cm³ | 90 s |

**Commentaries**

***Student A:*** This is almost a perfect table of results. There are three separate columns for the three sets of data collected. Each column has a heading and the units are given in the heading, although **s** is preferred to **secs** as the unit of time. The only mistake made is that the student has forgotten to convert the 1 minute to seconds for the second set of results obtained. **(2 marks scored – 1 for three columns with appropriate headings; 1 for correct units indicated. The third mark is for correct entering of all the data. This candidate fails to score this mark.)**

***Student B:*** Another good table of results. This time the student has decided to include the units in the columns rather than in the headings. This is perfectly acceptable, but unfortunately the student has forgotten to include the units of time for the fourth set of data obtained. **(2 marks scored – 1 for columns with headings; 1 for all data entered correctly.)**

## Worked Example 2

**Question**

A teacher investigates how the rate of reaction between magnesium and excess sulfuric acid changes as the concentration of the acid changes. The word equation for the reaction is:

magnesium + sulfuric acid → magnesium sulfate + hydrogen

The method she follows is:

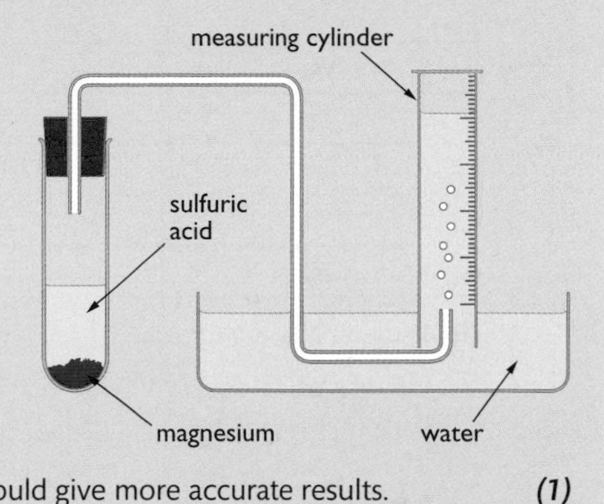


- add concentrated sulfuric acid to water to make acid of the required concentration
- use a measuring cylinder to pour 25 $cm^3$ of the diluted acid into a boiling-tube
- add magnesium chips to the boiling-tube and collect the gas produced as shown
- measure the volume of gas collected after 20 seconds.

The measuring cylinder is calibrated in cubic centimetres ($cm^3$).

***a)*** State one change that could be made to the apparatus that would give more accurate results. *(1)*

***b)*** The diagram shows the level of water in the measuring cylinder after one run. What volume of gas has been collected? *(1)*

***c)*** On which property of hydrogen does this method of gas collection depend? *(1)*

***d)*** The teacher notices that the boiling-tube felt hot after the reaction. She repeats the experiment and uses a thermometer to measure the temperature change of the reaction mixture.

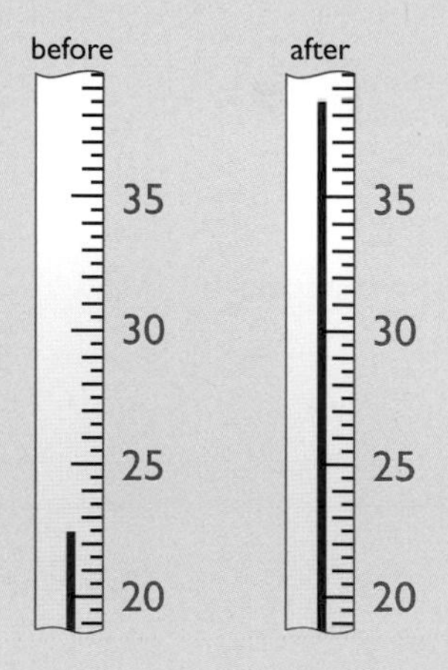


**(i)** The diagrams show the thermometer readings before and after the reaction. The thermometer is calibrated in degrees Celsius. What are the temperatures before and after mixing? *(2)*

**(ii)** Calculate the temperature change. *(1)*

**(iii)** What could be done to keep the temperature of the reaction mixture more constant? *(1)*

***e)*** State two variables, other than temperature, that must be kept constant to make the investigation a fair test. *(2)*

*f)* The table shows the teacher's results.

| Concentration of sulfuric acid (%) | Volume of gas ($cm^3$) collected in 20 seconds | | |
|---|---|---|---|
| | Run 1 | Run 2 | Run 3 |
| 10 | 46 | 48 | 47 |
| 15 | 62 | 63 | 62 |
| 20 | 75 | 74 | 71 |
| 30 | 65 | 63 | 67 |
| 40 | 50 | 33 | 46 |
| 50 | 33 | 34 | 35 |
| 60 | 27 | 23 | 22 |

(i) For which concentration of acid are the results most repeatable? *(1)*

(ii) Identify which one of the results is anomalous and explain what may have happened to cause this anomaly. *(2)*

(iii) Calculate the mean volume of gas collected in 20 seconds when the acid had a concentration of 60%. *(1)*

**Answers – part a**

***Student A:*** Use a pipette to measure the volume of the acid.

***Student B:*** Collect the gas in a gas syringe.

**Commentaries – part a**

***Student A:*** A pipette is more accurate than a measuring cylinder. **(1 mark scored)**

***Student B:*** A gas syringe is more accurate than a measuring cylinder. **(1 mark scored)**

A burette would also be more accurate.

A third way of scoring this mark would have been to use a divided reaction flask that keeps the magnesium and acid separate until the bung is replaced and the flask is shaken to mix the reactants.

**Answers – part b**

***Student A:*** 37 $cm^3$

***Student B:*** 43 $cm^3$

**Commentaries – part b**

***Student A:*** Correct answer. **(1 mark scored)**

***Student B:*** This student has not realised that the measuring cylinder is upside down and, therefore, that the scale starts from zero at the top. **(no marks)**

**Answers – part c**

***Student A:*** It does not dissolve in water.

***Student B:*** It is less dense than water.

**Commentaries – part c**

***Student A:*** Correct. **(1 mark scored)**

***Student B:*** This is a correct statement, but all gases are less dense than water and this is therefore not the property of hydrogen, in particular, that is important here. **(no marks)**

If hydrogen is collected in air, then the density is important. It would be collected by **upward** delivery since it is **less dense** than air.

**Answers – part d**

*Student A:* **(i)** before: 22.5°C; after 38.5°C. **(ii)** 16.0°C. **(iii)** Put the reaction tube in a water bath.

*Student B:* **(i)** before: 22.5; after 38.5. **(ii)** 16. **(iii)** Insulate the reaction tube by putting lagging around it.

**Commentaries – part d**

*Student A:* Answers to all parts are correct. **(4 marks scored)**

*Student B:* **(i)** Answers are correct. Although the units are not given, this is overlooked since the units are given in the question. **(ii)** An answer to two significant figures is acceptable here. Again the lack of units is overlooked. **(iii)** Lagging the test tube would make matters worse since less heat would be able to escape. The reaction mixture would, therefore, get even hotter than before. **(3 marks scored)**

**EXAMINER'S TIP**

In this experiment the same amount and the same mass are identical answers, since 1.2 g of magnesium is 0.05 mol, and both contain the same number of atoms of magnesium. You must be careful, however, when different substances are used in the experiments. If, for example, five different metals were being added to solutions of the same acid, then to make a fair test it is important to take the same **amount** of each metal in order to have the same number of atoms of each. The same mass would give different numbers of atoms.

**Answers – part e**

*Student A:* Same mass of magnesium. Same volume of acid.

*Student B:* Same amount of magnesium. Same size of magnesium chips.

**Commentaries – part e**

*Student A:* Same mass of magnesium is correct. Same volume of acid is not correct. As long as there is enough acid to completely cover the magnesium and react with it for the length of the experiment, the volume of acid used will make no difference to the volume of gas given off in 20 seconds. **(1 mark scored)**

*Student B:* Same amount (i.e. same number of moles) is correct. Same size of chips is correct. **(2 marks scored)**

**Answers – part f**

*Student A:* **(i)** 15% **(ii)** 33; The bung was not placed tightly enough into the reaction tube so some of the gas escaped and was not collected in the measuring cylinder. **(iii)** $\frac{27 + 23 + 22}{3} = 24\ cm^3$.

*Student B:* **(i)** second one **(ii)** 33 for the 40% acid; it was timed incorrectly **(iii)** 24.

**EXAMINER'S TIP**

When two or more marks are allocated for a calculation it is always advisable to show your working. If you make a mistake in calculating your final answer you can be given some marks for the correct method of working out your answer.

**Commentaries – part f**

*Student A:* **(i)** Correct answer since the three measurements of volume are closer together than are any of the others. **(1 mark scored)**

**(ii)** There are two measurements of 33 and this student has not stated which one, hence this mark is not scored. However, the reason given is valid for the 33 of the fifth experiment. In this situation the examiner would give the student the benefit of the doubt and assume that the experiment referred to is the one with 40% acid. **(1 mark scored)**

**(iii)** Correct answer. **(1 mark scored)**

*Student B:* **(i)** Correct answer. **(1 mark scored)**

**(ii)** The correct anomalous result has been identified, but the reason given cannot be credited since the student has not stated whether the time was too short or too long. **(1 mark scored)**

**(iii)** Correct answer. Only one mark is allocated for this answer, so working is not required. Also, the lack of units is overlooked since they were given in the question. **(1 mark scored)**

## Examination Questions

**1** The diagrams show a selection of apparatus you can find in a chemistry laboratory.

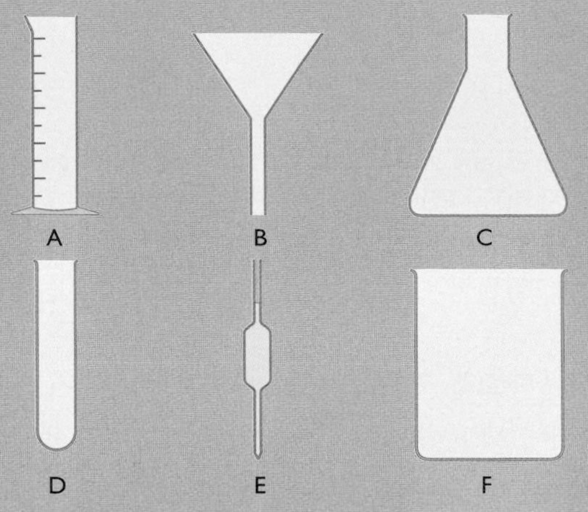


*a)* Complete the table by adding the name of each piece of apparatus. *(3)*

| Letter | Name |
|---|---|
| A | |
| B | filter funnel |
| C | |
| D | test tube |
| E | pipette |
| F | |

*b)* Select the **letters** of two pieces of apparatus that you would normally use to measure accurately the volume of a liquid. *(2)*

*c)* Which piece of apparatus is needed to separate particles of a solid from a liquid? *(1)*

***(Total 6 marks)***

**2** The diagram shows pieces of apparatus used to measure the volume of a liquid.

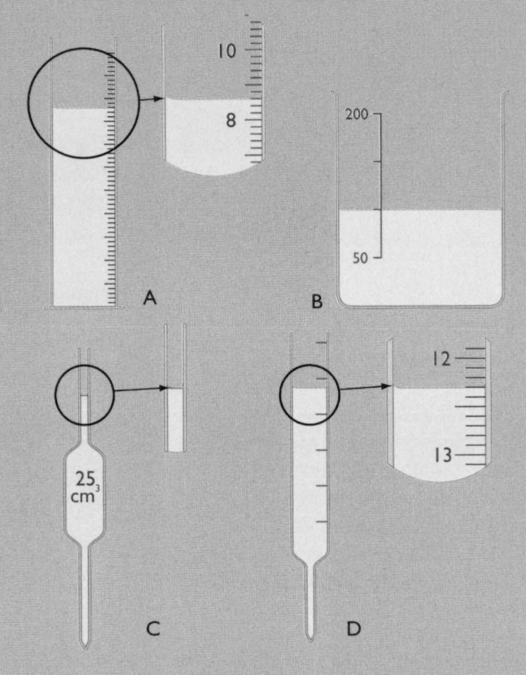


*a)* In the table write the name of each piece of apparatus and the volume of liquid being measured. *(8)*

| | Name of apparatus | Volume of liquid ($cm^3$) |
|---|---|---|
| A | | |
| B | | |
| C | | |
| D | | |

*b)* **(i)** Give the letter of the apparatus that is the **least** accurate. *(1)*

**(ii)** Give the letter of the apparatus that is the **most** accurate for measuring 20 $cm^3$ of a liquid. *(1)*

***(Total 10 marks)***

**3** Alcohols are flammable and can be used as fuels.

A student carried out an investigation to see if there was a relationship between the number of carbon atoms in an alcohol and how much energy it gave out when burned. The diagram shows the apparatus used.

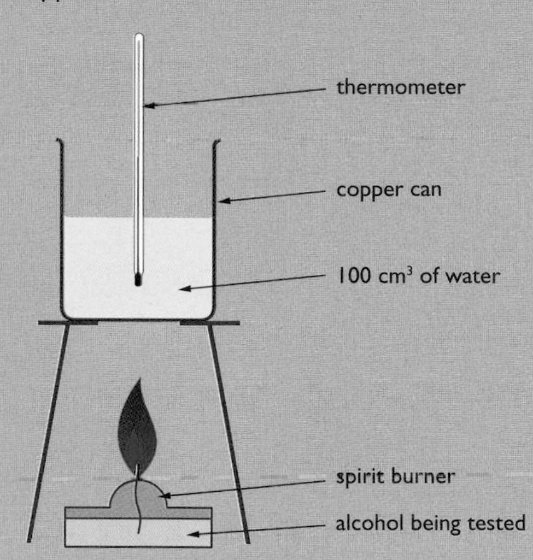


The student placed a spirit burner containing methanol under the can of water. She lit the spirit burner, heated the water for two minutes and put the spirit burner out. She repeated the experiment two more times. As the fuel was burned, the mass of the spirit burner became less. She repeated the experiment with three other alcohols.

*a)* The table shows the results obtained.

| Name of alcohol | Formula of alcohol | Mass of fuel used (g) | Temp change of water (°C) | Temp change per gram of fuel (°C g$^{-1}$) | Mean temp change per gram of fuel (°C g$^{-1}$) |
|---|---|---|---|---|---|
| methanol | $CH_3OH$ | 0.84 | | 48.2 | |
| | | 0.79 | 38.5 | 48.7 | |
| | | 0.76 | 37.0 | 48.7 | |
| ethanol | $C_2H_5OH$ | 0.78 | 52.5 | | |
| | | 0.64 | 43.0 | | |
| | | 0.68 | 45.5 | | |
| propanol | $C_3H_7OH$ | 0.54 | 37.0 | 68.5 | |
| | | 0.49 | 30.0 | 61.2 | |
| | | 0.57 | 46.5 | 81.6 | |
| butanol | $C_4H_9OH$ | 0.43 | 35.5 | 82.6 | |
| | | 0.47 | 38.5 | 81.9 | |
| | | 0.51 | 42.0 | 82.4 | |

(i) The diagrams show the thermometer readings before and after heating the water in the first experiment for methanol. Record the temperature shown on each thermometer. Calculate the **temperature change** for this experiment. *(3)*

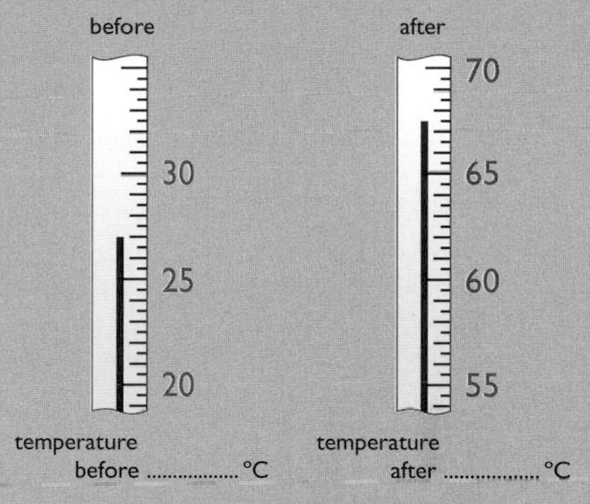


temperature before .................. °C    temperature after .................. °C

(ii) Other than measuring the temperature of the water before **and** after heating, what measurements must have been taken to get the results shown in the table above? *(1)*

(iii) The temperature change per gram of fuel used is calculated using the equation:

$$\text{temperature change per gram of fuel} = \frac{\text{temperature change}}{\text{mass of fuel used}}$$

Complete the table above to show the temperature change per gram of fuel for each experiment using ethanol. *(3)*

(iv) For each fuel, calculate the mean temperature change per gram of fuel. Record your answers in the table above. *(2)*

*b)* Use the information in the table above to help you answer this question.

(i) Are the results obtained for **methanol** reliable? Explain your answer. *(1)*

(ii) The results for **propanol** are not reliable. Explain why not. *(1)*

(iii) What should the student have done about the results for **propanol**? *(2)*

*c)* The student made the following conclusion. As the number of carbon atoms in any fuel increases, the energy given out when one gram of the fuel is burned also increases. Are the results obtained sufficient to support this conclusion? Explain your answer. *(2)*

***(Total 15 marks)***

**4** Solutions of lead(II) nitrate and potassium iodide react together to make the insoluble substance lead(II) iodide. The equation for the reaction is:

$Pb(NO_3)_2(aq) + 2KI(aq) \rightarrow 2KNO_3(aq) + PbI_2(s)$

An investigation was carried out to find how much precipitate formed with different volumes of lead(II) nitrate solution.

- A student measured out 15 $cm^3$ of potassium iodide solution using a measuring cylinder.
- He placed this solution in a clean boiling-tube.
- Using a clean measuring cylinder, he measured out 2 $cm^3$ of lead(II) nitrate solution (of the same concentration, in mol/$dm^3$, as the potassium iodide solution). He added this to the potassium iodide solution.
- A cloudy yellow mixture formed and this was left to settle.
- The student then measured the height (in cm) of the precipitate using a ruler.

The student repeated the experiment using different volumes of lead(II) nitrate solution. The graph shows the results obtained.

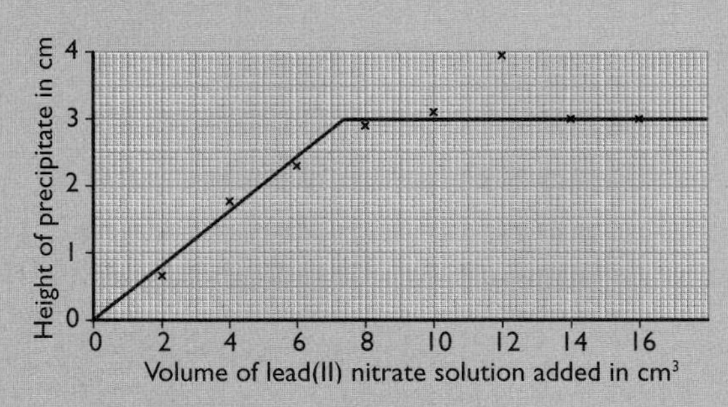


*a)* **(i)** On the graph, circle the point which seems to be anomalous. *(1)*

**(ii)** Explain **two** things that the student may have done in the experiment to give this anomalous result. *(4)*

**(iii)** Why must the graph line go through (0,0)? *(1)*

*b)* Suggest a reason why the height of the precipitate stops increasing. *(1)*

*c)* **(i)** How much precipitate, in $cm^3$, has been made in the tube? *(1)*

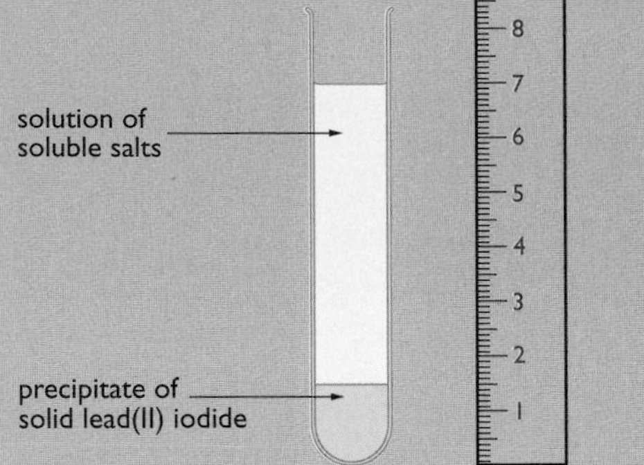


**(ii)** Use the graph to find the volume of lead(II) nitrate solution needed to make this amount of precipitate. *(1)*

*d)* After he had plotted the graph, the student decided he should obtain some more results.

**(i)** Suggest what volumes of lead(II) nitrate solution he should use. *(1)*

**(ii)** Explain why he should use these volumes. *(1)*

*e)* Suggest a different method for measuring the amount of precipitate formed. This method **must not** be based on the height of the precipitate. *(4)*

***(Total 15 marks)***

**5** Sodium thiosulfate solution and hydrochloric acid react to form a precipitate of sulfur. This precipitate makes the mixture go cloudy.

- A student placed 10 $cm^3$ of sodium thiosulfate solution and 30 $cm^3$ of water in a conical flask.
- She then added 10 $cm^3$ of hydrochloric acid.
- She placed the conical flask on a piece of paper with a black cross.
- She timed how long it took until she could not see the cross through the conical flask.

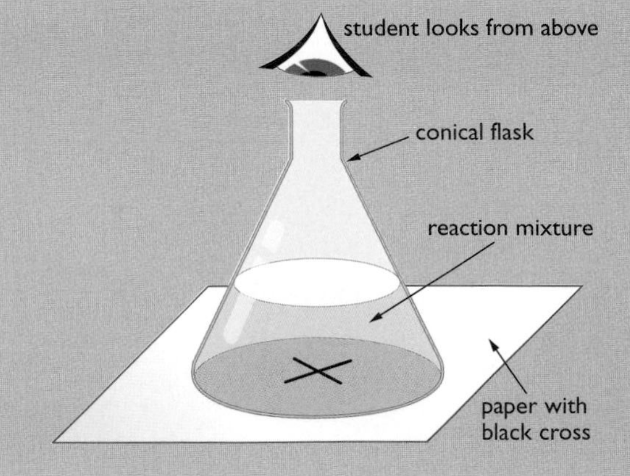

She repeated the experiment using the same volumes of sodium thiosulfate solution, water and hydrochloric acid at different temperatures.

The graph shows her results.

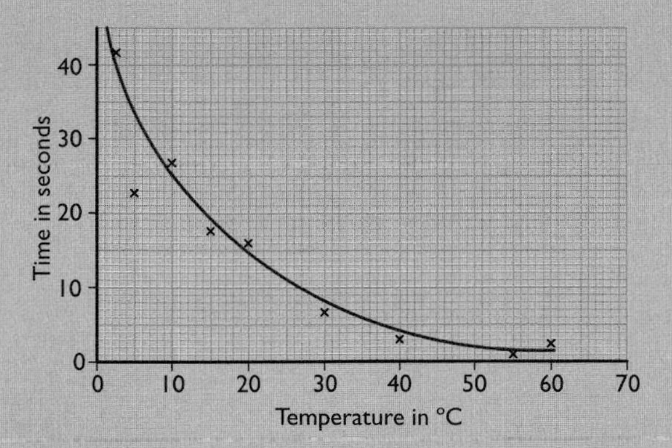


*a)* When should the student have started her stopwatch? *(1)*

*b)* (i) Circle on the graph one result that is anomalous. *(1)*

(ii) Explain what may have happened during the experiment to produce the anomalous result. *(3)*

(iii) Use the graph to find the time taken for the cross to be no longer visible at 32°C. *(1)*

*c)* The student used her results to work out the rate of the reaction at different temperatures. The equation she used was:

$$\text{rate of reaction} = \frac{1}{\text{time taken}}$$

Calculate the rate of reaction at 32°C. *(2)*

*d)* The second graph shows how the rate of reaction changed as the temperature was increased.

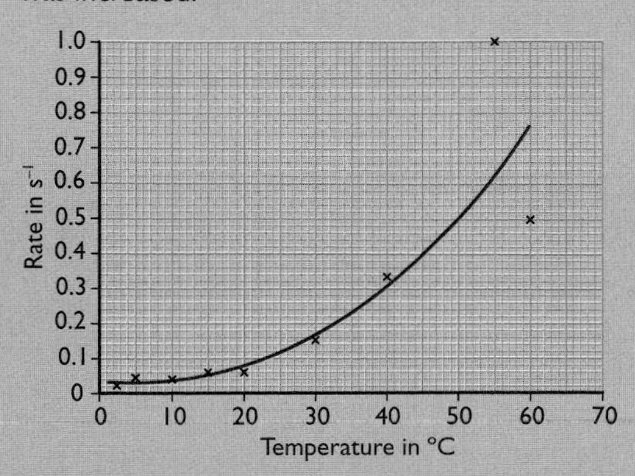


The student looked at her graph and decided the results were least accurate at high temperatures. Give **two** reasons why the results are least accurate at high temperatures. *(2)*

*e)* (i) Describe the relationship between the temperature and the rate of reaction. *(2)*

(ii) Use scientific knowledge to explain why increasing the temperature has this effect on the rate of reaction. *(3)*

*f)* The student decided to use the same reaction to investigate how the rate of reaction was affected by changing the volume of hydrochloric acid. Outline the method she should use. *(4)*

***(Total 19 marks)***

**6** Sulfur dioxide is a toxic and acidic gas. When it is dissolved in rainwater, acid rain is formed. A student wanted to investigate how the solubility of sulfur dioxide varies with temperature. He suggested the following plan.

- Place 100 $cm^3$ of water in a beaker.
- Place the beaker on a balance.
- Bubble gas into the water using the apparatus shown in the diagram.

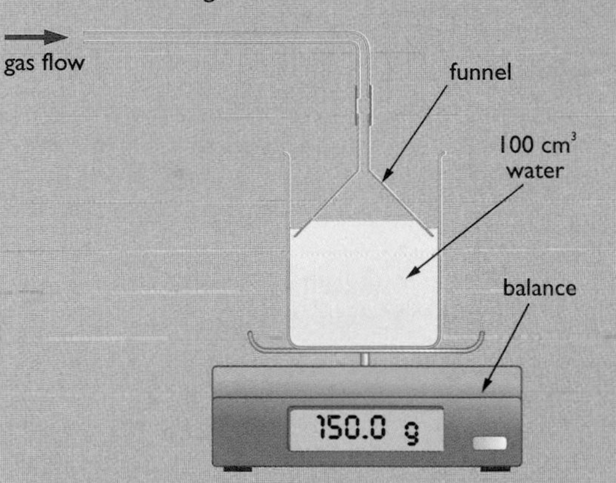


- Continue to pass gas into the water until the mass of the beaker and its contents stops changing.
- Repeat the experiment at different temperatures.

The table shows the student's preliminary results.

| Mass of beaker and water at start (g) | Temperature (°C) | Mass of beaker and solution at end (g) |
|---|---|---|
| 150.0 | 15 | 162.7 |
| 150.0 | 20 | 160.6 |
| 150.0 | 25 | 159.0 |
| 150.0 | 30 | 157.6 |

*a)* What mass of sulfur dioxide dissolved in the water at 20°C? *(1)*

*b)* How does the solubility of sulfur dioxide change as the temperature is increased? *(1)*

*c)* State **one** safety precaution the student should take when doing this experiment. Why is this precaution needed? *(2)*

*d)* The student did an experiment at a temperature just over 30°C. He noticed that the balance reading increased at first, but then slowly decreased and did not become constant. He did an experiment at about 90°C. The decrease in the balance reading occurred much more quickly than at 30°C.

(i) Suggest why the mass decreased slowly when the temperature was just over 30°C. *(1)*

(ii) Suggest why the mass decreased more quickly at about 90°C. *(1)*

*e)* Sulfur dioxide is an acidic gas. Outline another way the student could compare the amount of sulfur dioxide dissolved in the water at different temperatures. *(1)*

*(Total 7 marks)*

**7** Barium chloride and silver nitrate solutions can be used to test for the presence of some anions in solution. The table shows the results obtained when some solutions are tested with these reagents.

| Solution | 1. $BaCl_2$(aq)<br>2. then HCl(aq) | 1. $AgNO_3$(aq)<br>2. then $HNO_3$(aq) |
|---|---|---|
| sodium nitrate | 1. no change<br>2. no change | 1. no change<br>2. no change |
| sodium chloride | 1. no change<br>2. no change | 1. white precipitate<br>2. no change |
| sodium sulfate | 1. white precipitate<br>2. no change | 1. no change<br>2. no change |
| sodium carbonate | 1. white precipitate<br>2. disappears, fizzing | 1. white precipitate<br>2. disappears, fizzing |

A student is given six solutions, labelled **U**, **V**, **W**, **X**, **Y** and **Z**, each containing a mixture of **two** sodium compounds from the table above. He tests each solution with the reagents shown.

The student is then told to use his results and the information from the table above to identify the two anions present. His results and conclusions are shown below.

| Solution | 1. $BaCl_2$(aq)<br>2. then HCl(aq) | 1. $AgNO_3$(aq)<br>2. then $HNO_3$(aq) | Anions present |
|---|---|---|---|
| **U** | 1. white precipitate<br>2. precipitate disappears, fizzing | 1. white precipitate<br>2. precipitate disappears, fizzing | carbonate nitrate |
| **V** | 1. white precipitate<br>2. no change | 1. white precipitate<br>2. no change | chloride sulfate |
| **W** | 1. white precipitate<br>2. no change | 1. no change<br>2. no change | chloride nitrate |
| **X** | 1. no change<br>2. no change | 1. white precipitate<br>2. no change | carbonate sulfate |
| **Y** | 1. white precipitate<br>2. fizzing | 1. white precipitate<br>2. precipitate disappears, fizzing | carbonate chloride |
| **Z** | 1. white precipitate<br>2. precipitate disappears, fizzing | 1. white precipitate<br>2. precipitate remains, fizzing | nitrate sulfate |

*a)* Identify, choosing from the letters **U**, **V**, **W**, **X**, **Y** and **Z**, **two** solutions in which **both** anions have been **correctly** identified. *(2)*

*b)* Identify, choosing from the letters **U**, **V**, **W**, **X**, **Y** and **Z**, **two** solutions in which **both** anions have been **wrongly** identified. *(2)*

*(Total 4 marks)*

A number of examiner's tips and advice have already been given throughout Sections A to F.

Section F has provided information and advice on answering questions based on investigative skills (AO3).

The purpose of this section is to offer some advice on answering questions assessing skills AO1 and AO2. This will, once again, be attempted by providing different answers to some International GCSE examination-style questions and then discussing the relative merits of each answer.

# Vocabulary used by examiners

First, it would be useful to have some understanding of the 'command words' used by examiners when writing questions. The table below provides information on the meaning of some specific command words used in examinations:

| Command word | Meaning |
| --- | --- |
| analyse and interpret | Identify, with reasons, the essential features of the information or data given. This may involve some manipulation of the data. |
| calculate | Perform a numerical calculation using the data supplied and supply a final answer to the appropriate degree of precision. You are advised to show all of you working, since some marks may be awarded for a correct method, even if arithmetical mistakes are made. |
| describe | Requires candidates to state in words (using diagrams where appropriate) the main points of the item or process for which the description is requested. The amount of description required should be interpreted in the light of the marks available. |
| distinguish | Identify appropriate differences in a given context |
| explain | Use appropriate scientific knowledge to give reasons or explanations for the information or data given. This will usually be a 2, 3 or 4 mark question and answers should go beyond just repetition or reorganisation of the information or data provided. You should check that your response answers the question, 'Why ....?' |
| identify | Short factual answers are required. If asked to identify a substance or a particle, then either the name or the correct formula may be given. |
| name, state, give | Short, factual answers are required, possibly with precise use of scientific terms. Often one-word answers are sufficient. |
| predict | Implies that candidates are not expected to produce the required answer by recall but by making a logical connection between other pieces of information. Such information may be wholly given in the question or may depend on answers extracted in an earlier part of the question. 'Predict' also implies a concise answer with no supporting statement required. |
| suggest | The answer required may include material or ideas that have not been learnt directly from the specification. A reasonable suggestion, using **relevant** scientific knowledge and understanding of related topics, is required. The question may be related to unfamiliar situations or may relate to familiar situations in which there may be more than one acceptable answer. |
| using the information in the diagram/on the graph/in the table | Refer only to the information presented in the question, **not** other examples or knowledge. |

# Examination-style questions and answers

## Worked Example 1

### Question

Rubidium reacts with oxygen, chlorine and water in a similar way to other Group 1 elements.

(i) Suggest the formula of the compound formed when rubidium reacts with oxygen and with chlorine. *(2)*

(ii) A small piece of rubidium is added to water in a glass beaker. Suggest two observations you are likely to make during the reaction. *(2)*

(iii) Write a balanced chemical equation for the reaction between rubidium and water. Include state symbols. *(3)*

### Answers

***Student A:*** (i) $Rb_2O$ and $RbCl$

(ii) Effervescence and rubidium moves around the surface of the water

(iii) $2Rb(s) + H_2O(l) \rightarrow Rb_2O(aq) + H_2(g)$

***Student B:*** (i) $RbCl$ and $Rb_2O$

(ii) Bubbles of gas, effervescence, a flame is seen and rubidium sinks

(iii) $2Rb(s) + 2H_2O(l) \rightarrow 2RbOH(s) + H_2(g)$

### Commentaries

***Student A:*** (i) Both formulae are correct. **(2 marks scored)**

(ii) Both observations are reasonable, since that is what happens when lithium, sodium or potassium is added to water. **(2 marks scored)**

(iii) The formula for the rubidium compound formed is incorrect. It is a very common mistake to assume that a metal oxide, and not a metal hydroxide, is formed when a metal reacts with water. Because the formula for one of the substances in the equation is incorrect, it is usual to give zero for this answer; the balancing and state symbol marks are not awarded if any one of the formulae is incorrect. However, in this particular case, $Rb_2O$ is not an unreasonable suggestion for the compound formed, and hence the mark for state symbols might be given. **(1 mark scored)**

***Student B:*** (i) Both formulae are correct. Although they are given in the wrong order, this does not matter unless specific spaces have been left for each answer, in which case the student would have lost 1 mark out of the 2. **(2 marks scored)**

(ii) Bubbles of gas is correct, but effervescence is an alternative to bubbles and does not score an extra mark. 'A flame is seen' is a reasonable observation to suggest, since a flame is produced with potassium, and rubidium is more reactive than potassium since it is below it in the group. However, this student has added another observation than is not reasonable. There is no evidence to suggest that rubidium will sink; the expectation is that it will do the same as the other alkali metals above it in the group and float on water. Hence this has to count as an incorrect observation and cancels out one of the correct ones. **(1 mark scored)**

(iii) All formulae are correct and the equation is balanced. However, the state symbol for RbOH should be (aq) not (s), so that state symbol mark is lost. **(2 marks scored)**

## Worked Example 2

### Question

A student has four solutions labelled **A**, **B**, **C** and **D**.

Each solution contains one compound from the following list:

$KNO_3$  $FeCl_2$  $Fe(NO_3)_3$  $CuSO_4$  $NH_4Cl$

The student did some simple tests to identify the compounds present.

The table shows the tests and observations.

| Solution | Colour | Add sodium hydroxide solution | Add dilute nitric acid and silver nitrate solution |
|---|---|---|---|
| **A** | colourless | pungent gas given off | white precipitate |
| **B** | blue | blue precipitate | no change |
| **C** | colourless | no change | no change |
| **D** | green | green precipitate | white precipitate |

*a)* (i) What is the pungent gas formed by solution **A**? *(1)*

(ii) Which ion must be present in **A** for the white precipitate to form? *(1)*

(iii) Which ion must be present in **B** for the blue precipitate to form? *(1)*

(iv) Which ion must be present in **D** for the green precipitate to form? *(1)*

*b)* (i) Which compound in the list can be indentified using barium chloride solution? *(1)*

(ii) State one compound in the list that can be identified using a flame test. State the colour of the flame. *(2)*

*c)* (i) Silver nitrate solution, $AgNO_3(aq)$, is added to a solution of lithium iodide, LiI. Describe what is seen. *(1)*

(ii) Write the chemical equation, including state symbols, for the reaction. *(3)*

### Answers

***Student A:*** ***a)*** **(i)** $NH_3$ **(ii)** $Cl^-$ **(iii)** $Cu^{2+}$ **(iv)** $Fe^{2+}$

***b)*** **(i)** $CuSO_4$ **(ii)** $KNO_3$; Lilac

***c)*** **(i)** Yellow precipitate **(ii)** $Ag^+(aq) + I^-(aq) \rightarrow AgI(s)$

***Student B:*** ***a)*** **(i)** ammonium, $NH_3$ **(ii)** chlorine ion **(iii)** copper ion **(iv)** iron ion

***b)*** **(i)** sulfate ion **(ii)** copper sulfate; blue-green

***c)*** **(i)** Yellow solid **(ii)** $LiI(aq) + AgNO_3(aq) \rightarrow LiNO_3(s) + AgI(aq)$

### Commentaries

***Student A:*** ***a)*** All four marks scored. Formulae are acceptable for the answers since the question did not ask for names.

***b)*** All three marks scored.

***c)*** **(i)** Yellow precipitate is correct **(1 mark scored)**

**(ii)** Although an ionic equation was not asked for it is perfectly acceptable. All formulae are correct; the equation is balanced and all state symbols are correct. **(3 marks scored)**

***Student B:*** ***a)*** **(i)** Although the correct formula for the gas is given, the mark is lost because of an incorrect name. It is a very common mistake to confuse the names of the gas (ammonia) and the ammonium ion ($NH_4^+$), so take care when writing your answers.

(ii) The correct name for the ion is chlori**d**e, not chlori**n**e, so this answer does not score.

(iii) The correct answer here is copper(II), since copper can also form a copper(I) ion. However, in this case copper would be marked correct, since only one copper ion is mentioned in the specification. **(1 mark scored)**

(iv) Iron ion is not sufficient to score the mark. Iron forms two ions: iron(II) and iron(III). Iron(II) ions produce a green precipitate with sodium hydroxide solution, whilst iron(III) ions produce a brown precipitate.

*b)* (i) The student has not answered the question. Although the sulfate ion can be identified using barium chloride solution, the question asked for the name of the **compound** that could be identified.

(ii) Both marks are scored here. Although the identification of copper(II) ions using a flame test is not on the specification, it nevertheless can be identified this way, and the colour of the flame is correct. **(2 marks scored)**

*c)* (i) Solid is an acceptable alternative to precipitate. **(1 mark scored)**

(ii) All formula are correct and the equation is correctly balanced. However, the state symbols for $LiNO_3$ and AgI are the wrong way around. **(2 marks scored)**

## Worked Example 3

**Question**

The conversion of ethane to ethene is represented by the following equation:

$C_2H_6(g) \rightarrow C_2H_4(g) + H_2(g)$

The table shows some bond energies

| Bond | C–C | C=C | C–H | H–H |
|---|---|---|---|---|
| Bond energy in kJ/mol | 348 | 612 | 412 | 436 |

*a)* (i) Ethane and hydrogen contain only single bonds. Ethene contains both single and double bonds. Draw a displayed formula for each of the molecules ethane and ethene in the following equation.

ethane → ethene + hydrogen

H–H *(2)*

(ii) Use your displayed formulae and the information in the table to calculate the energy change occurring in the conversion of ethane into ethene. *(3)*

*b)* At room temperature and pressure the conversion of ethane to ethene is very slow. State **two** changes in the conditions that would increase the rate of this reaction. *(2)*

*c)* The equation below represents a reaction of ethene that is used in industry.

$2C_2H_4(g) + O_2(g) \rightleftharpoons 2(CH_2)_2O(g) \quad \Delta H = -214$ kJ/mol

(i) What do the symbols $\rightleftharpoons$ and ΔH represent? *(2)*

(ii) The reaction is carried out at a pressure of 2 atmospheres and a temperature of 300°C.

Predict **and** explain what would happen to the amount of product formed at equilibrium if, first of all, the pressure was increased at constant temperature and, secondly, the temperature was increased at constant pressure. *(4)*

**Answers**

*Student A: a)* (i)

```
   H  H              H  H
   |  |              |  |
H—C—C—H     and      C=C
   |  |              |  |
   H  H              H  H
```

(ii) $\Sigma$ (bonds broken) = 1(C−C) + 6(C−H) = 348 + (6 × 412) = 2820 kJ

$\Sigma$ (bonds made) = 1(C=C) + 4(C−H) + 1(H−H)

= 612 + (4 × 412) + 436 = 2696 kJ

$\Delta H$ = (2696 − 2820) = **−124 kJ/mol**

*b)* Increasing the temperature and increasing the pressure.

*c)* (i) $\rightleftharpoons$ represents a reversible reaction

$\Delta H$ represents the enthalpy change in the reaction

(ii) An increase in pressure would cause the amount of product to increase. This is because there are fewer molecules on the right hand side and hence the equilibrium position will shift to the right.

An increase in temperature would cause the amount of product to decrease. This is because the forward reaction is exothermic and hence the position of equilibrium will shift in the endothermic direction, which is to the left.

*Student B:* *a)* (i)

```
    H   H              H   H
    |   |              |   |
H — C — C — H   and    C = C
    |   |              |   |
    H   H              H   H
```

(ii) $\Sigma$ (bonds broken) = 1(C−C) + 2(C−H) = 348 + (2 x 412) = 1172 kJ

$\Sigma$ (bonds made) = 1(C=C) + 1(H−H) = 612 + 436 = 1048 kJ

$\Delta H$ = (1172 − 1048) = **124 kJ**

*b)* Increase in pressure and increase in concentration of reactants.

*c)* (i) $\rightleftharpoons$ represents that the reaction can go in both directions

$\Delta H$ represents the heat of reaction

(ii) There are more molecules of gas on the left hand side so the equilibrium will shift to the right when the pressure is increased. This will increase the amount of product formed.

The equilibrium will shift in the endothermic direction when the temperature increases, which is to the right in this reaction. This will increase the amount of product formed.

**Commentaries**

*Student A:* *a)* (i) Both displayed formulae are correct. **(2 marks scored)**

(ii) Calculation of bonds broken and bonds made are both correct **(2 marks scored)**. However, the student has calculated $\Delta H$ incorrectly. It should be $\Sigma$ (bonds broken) − $\Sigma$ (bonds made) to give an answer of +124 kJ/mol. Therefore the third mark is lost.

*b)* The student has correctly indentified that an **increase** in both temperature and pressure will cause an increase in the rate of reaction. (**2 marks scored)**

*c)* (i) The explanation of both symbols is correct. **(2 marks scored)**

(ii) The prediction and explanation are correct for an increase in pressure. **(2 marks scored)**

Although the student has calculated $\Delta H$ incorrectly in *a)* **(ii)**, the prediction and explanation offered are consistent with the value calculated. Hence the answer is marked correct under the 'error carried forward' rule. **(2 marks scored)**

An increase in pressure causes an increase in rate because the reactant is a gas. A change in pressure has no effect on reactions where the reactants are either solids or liquids.

*Student B:* *a)* **(i)** Both displayed formulae are correct. **(2 marks scored)**

**(ii)** The calculation is correct. The omission of the positive sign is ignored. However, if the reaction had been exothermic, it would have been **essential** to include the negative sign for $\Delta H$. **(3 marks scored)**

This student decided to use **only** those bonds that are actually broken and formed in the reaction. This is perfectly acceptable.

*b)* Increase in pressure is correct and scores a mark, but no extra mark can be given for increase in concentration of reactant. Since the reactant is a gas, an increase in pressure causes an increase in concentration, so these two answers are the same. **(1 mark scored)**

*c)* **(i)** The reaction can go in both directions is equivalent to the reaction is reversible.

The student should have said heat **change** in the reaction, but this would be ignored. **(2 marks scored)**

**(ii)** More molecules on the left hand side is equivalent to fewer molecules of the right. Hence the explanation and the prediction for an increase in pressure are both correct. **(2 marks scored)**

Explanation and prediction are again both correct. **(2 marks)**

Group

| Period | 1 | 2 | | | | | | | | | | | 3 | 4 | 5 | 6 | 7 | 0 |
|---|---|---|---|---|---|---|---|---|---|---|---|---|---|---|---|---|---|---|
| 1 | | | | | | | | 1<br>H<br>Hydrogen<br>1 | | | | | | | | | | 4<br>He<br>Helium<br>2 |
| 2 | 7<br>Li<br>Lithium<br>3 | 9<br>Be<br>Beryllium<br>4 | | | | | | | | | | | 11<br>B<br>Boron<br>5 | 12<br>C<br>Carbon<br>6 | 14<br>N<br>Nitrogen<br>7 | 16<br>O<br>Oxygen<br>8 | 19<br>F<br>Fluorine<br>9 | 20<br>Ne<br>Neon<br>10 |
| 3 | 23<br>Na<br>Sodium<br>11 | 24<br>Mg<br>Magnesium<br>12 | | | | | | | | | | | 27<br>Al<br>Aluminium<br>13 | 28<br>Si<br>Silicon<br>14 | 31<br>P<br>Phosphorus<br>15 | 32<br>S<br>Sulfur<br>16 | 35.5<br>Cl<br>Chlorine<br>17 | 40<br>Ar<br>Argon<br>18 |
| 4 | 39<br>K<br>Potassium<br>19 | 40<br>Ca<br>Calcium<br>20 | 45<br>Sc<br>Scandium<br>21 | 48<br>Ti<br>Titanium<br>22 | 51<br>V<br>Vanadium<br>23 | 52<br>Cr<br>Chromium<br>24 | 55<br>Mn<br>Manganese<br>25 | 56<br>Fe<br>Iron<br>26 | 59<br>Co<br>Cobalt<br>27 | 59<br>Ni<br>Nickel<br>28 | 63.5<br>Cu<br>Copper<br>29 | 65<br>Zn<br>Zinc<br>30 | 70<br>Ga<br>Gallium<br>31 | 73<br>Ge<br>Germanium<br>32 | 75<br>As<br>Arsenic<br>33 | 79<br>Se<br>Selenium<br>34 | 80<br>Br<br>Bromine<br>35 | 84<br>Kr<br>Krypton<br>36 |
| 5 | 85<br>Rb<br>Rubidium<br>37 | 88<br>Sr<br>Strontium<br>38 | 89<br>Y<br>Yttrium<br>39 | 91<br>Zr<br>Zirconium<br>40 | 93<br>Nb<br>Niobium<br>41 | 96<br>Mo<br>Molybdenum<br>42 | (99)<br>Tc<br>Technetium<br>43 | 101<br>Ru<br>Ruthenium<br>44 | 103<br>Rh<br>Rhodium<br>45 | 106<br>Pd<br>Palladium<br>46 | 108<br>Ag<br>Silver<br>47 | 112<br>Cd<br>Cadmium<br>48 | 115<br>In<br>Indium<br>49 | 119<br>Sn<br>Tin<br>50 | 122<br>Sb<br>Antimony<br>51 | 128<br>Te<br>Tellurium<br>52 | 127<br>I<br>Iodine<br>53 | 131<br>Xe<br>Xenon<br>54 |
| 6 | 133<br>Cs<br>Caesium<br>55 | 137<br>Ba<br>Barium<br>56 | 139 ■<br>La<br>Lanthanum<br>57 | 178<br>Hf<br>Hafnium<br>72 | 181<br>Ta<br>Tantalum<br>73 | 184<br>W<br>Tungsten<br>74 | 186<br>Re<br>Rhenium<br>75 | 190<br>Os<br>Osmium<br>76 | 192<br>Ir<br>Iridium<br>77 | 195<br>Pt<br>Platinum<br>78 | 197<br>Au<br>Gold<br>79 | 201<br>Hg<br>Mercury<br>80 | 204<br>Tl<br>Thallium<br>81 | 207<br>Pb<br>Lead<br>82 | 209<br>Bi<br>Bismuth<br>83 | (210)<br>Po<br>Polonium<br>84 | (210)<br>At<br>Astatine<br>85 | (222)<br>Rn<br>Radon<br>86 |
| 7 | (223)<br>Fr<br>Francium<br>87 | (226)<br>Ra<br>Radium<br>88 | (227) ■<br>Ac ■<br>Actinium<br>89 | | | | | | | | | | | | | | | |

| | | | | | | | | | | | | | | |
|---|---|---|---|---|---|---|---|---|---|---|---|---|---|---|
| ■ | 140<br>Ce<br>Cerium<br>58 | 141<br>Pr<br>Praseodymium<br>59 | 144<br>Nd<br>Neodymium<br>60 | (147)<br>Pm<br>Promethium<br>61 | 150<br>Sm<br>Samarium<br>62 | 152<br>Eu<br>Europium<br>63 | 157<br>Gd<br>Gadolinium<br>64 | 159<br>Tb<br>Terbium<br>65 | 163<br>Dy<br>Dysprosium<br>66 | 165<br>Ho<br>Holmium<br>67 | 167<br>Er<br>Erbium<br>68 | 169<br>Tm<br>Thulium<br>69 | 173<br>Yb<br>Ytterbium<br>70 | 175<br>Lu<br>Lutetium<br>71 |
| ■ ■ | 232<br>Th<br>Thorium<br>90 | (231)<br>Pa<br>Protoactinium<br>91 | 238<br>U<br>Uranium<br>92 | (237)<br>Np<br>Neptunium<br>93 | (242)<br>Pu<br>Plutonium<br>94 | (243)<br>Am<br>Americium<br>95 | (247)<br>Cm<br>Curium<br>96 | (247)<br>Bk<br>Berkelium<br>97 | (251)<br>Cf<br>Californium<br>98 | (254)<br>Es<br>Einsteinium<br>99 | (253)<br>Fm<br>Fermium<br>100 | (256)<br>Md<br>Mendelevium<br>101 | (254)<br>No<br>Nobelium<br>102 | (257)<br>Lr<br>Lawrencium<br>103 |

Key: a / X / Name / b

a = relative atomic mass
X = atomic symbol
b = atomic number

(Masses in parentheses are the mass numbers of the most stable isotope)

# Glossary

**acid** A substance that produces hydrogen ions ($H^+$) when dissolved in water.

**acid–alkali indicator** A substance used to show if a solution is acidic or alkaline. It will have a different colour in each type of solution.

**acidic solution** A solution that has a $pH < 7$.

**acid rain** Rain that has a $pH < 5$; produced when gases such as sulfur dioxide dissolve in rainwater.

**activation energy** The minimum energy required by colliding particles for reaction to occur.

**alkali** A base that dissolves in water to form hydroxide ions ($OH^-$).

**alkaline solution** A solution that has a $pH > 7$.

**alkane** A saturated hydrocarbon with the general formula $C_nH_{2n+2}$.

**alkene** An unsaturated hydrocarbon with the general formula $C_nH_{2n}$.

**anion** A negatively charged ion. For example, $Cl^-$, $O^{2-}$, $OH^-$, $NO_3^-$, etc.

**atom** The smallest particle of an element that can take part in a chemical reaction.

**atomic number** The number of protons in the nucleus of an atom.

**base** A substance that neutralises an acid to form a salt.

**bauxite** The main ore of aluminium, from which aluminium is extracted.

**brine** A concentrated solution of sodium chloride in water.

**catalyst** A substance that speeds up a chemical reaction but which is chemically unchanged at the end of the reaction.

**cation** A positively charged ion. For example, $Na^+$, $Mg^{2+}$, $Al^{3+}$, $NH_4^+$, etc.

**combustion** A chemical reaction involving burning.

**compound** A substance made up of two or more elements chemically combined together.

**contact process** The name of the industrial process used for making sulfuric acid.

**corrosion** A chemical reaction between a metal and oxygen in the air.

**covalent bond** The force of attraction between the nuclei of two atoms and a pair of electrons shared between them.

**cracking** The process of breaking long-chain alkane molecules into short-chain alkanes and alkenes.

**crude oil** A naturally occurring mixture of many hydrocarbons.

**cryolite** A mineral of aluminium used in its molten form for dissolving aluminium oxide in the electrolytic manufacture of aluminium.

**decomposition** Breaking down a compound into simpler substances (either elements and/or other compounds).

**diffusion** The movement of particles (atoms, molecules or ions) from an area of high concentration to one of lower concentration.

**displacement reaction** A reaction in which one substance replaces another, for example chlorine displacing bromine from a bromide or zinc displacing copper from a copper salt or zinc displacing copper from copper(II) oxide.

**distillation** The process of separating a liquid from a solution of a solid in the liquid.

**dynamic equilibrium** The condition that exists when the rate of the forward and backward reactions in a reversible reaction mixture are equal.

**effervescence** The rapid production of bubbles of gas produced during a reaction taking place in aqueous solution.

**electrode** A solid electrical conductor that forms the connection between the electrolyte and the external electrical circuit in electrolysis.

**electrolysis** The decomposition of a substance by passing an electric current through it.

**electronic configuration** The way that the electrons are arranged into shells in an atom, for example Na 2.8.1.

**electron** A negatively charged sub-atomic particle.

**element** A substance made up of atoms that all contain the same number of protons.

**empirical formula** The simplest whole-number ratio of atoms present in a compound.

**endothermic reaction** A reaction that takes in heat energy.

**exothermic reaction** A reaction that gives out heat energy.

**Faraday constant** The quantity of electric charge carried by one mole of electrons.

**fermentation** The conversion of glucose to ethanol and carbon dioxide.

**flame test** A method of identifying a metal cation by the colour it produces in a non-luminous Bunsen flame.

**fraction** A mixture containing several compounds all of which have similar boiling points.

**fractional distillation** The process of separating the liquids in a mixture of miscible liquids, using the fact that the liquids have different boiling points.

**Haber process** The name of the industrial process used for making ammonia.

**halogen** An element in Group 7 of the Periodic Table.

**homologous series** A group of organic compounds in which each member differs from the next member by a $-CH_2-$ unit.

**hydrocarbon** A compound made up of only the elements hydrogen and carbon.

**immiscible liquids** Liquids that do not mix together.

**intermolecular force** The force of attraction between individual molecules.

**ion** An electrically charged atom or group of atoms, for example $Na^+$, $Cl^-$, $NH_4^+$, $SO_4^{2-}$, etc.

**ionic bond** The force of attraction between oppositely charged ions in a compound.

**isomers** Compounds that have the same molecular formula but different displayed formulae.

**isotopes** Atoms of the same element that have different masses. They contain the same number of protons but different numbers of neutrons.

**litmus** The most common acid–alkali indicator. It is red in solutions with a pH $\leq 5$ and blue in solutions $\geq 8$.

**mass number** The sum of the number of protons and neutrons in the nucleus of an atom.

**metallic bond** The force of attraction between the positive metal ions and the delocalised electrons in a metal.

**mineral** A rock found in the Earth's crust that contains a metal or a metal compound.

**miscible liquids** Liquids that can mix together.

**mixture** A combination of two or more substances (elements and/or compounds) that are not chemically joined together.

**molar enthalpy change** The net heat energy change per mole in a chemical reaction.

**molar volume** The volume of one mole of a gas. At r.t.p. it is 24 $dm^3$.

**mole** The amount of substance containing $6 \times 10^{23}$ particles (atoms, molecules or formulae) of the substance.

**molecular formula** The exact number of atoms of each element present in a molecule of a substance.

**molecule** The smallest particle of a substance that can have a separate, independent existence.

**monatomic molecule** A molecule that contains only one atom (e.g. those of the noble gases).

**monomer** The unit molecule from which a polymer is made.

**neutralisation** The reaction between an acid and a base to produce a salt.

**neutral solution** A solution that has a pH of 7.

**neutron** A neutral sub-atomic particle present in the nucleus of atoms.

**ore** A mineral that contains enough of a metal or a metal compound to make it worthwhile extracting the metal.

**oxidation** The removal of electrons from a substance, or the addition of oxygen.

**oxidising agent** A substance that is capable of oxidising another substance.

**Periodic Table** A table in which the elements are arranged in order of increasing atomic number. Elements with similar properties appear in the same groups (vertical columns).

**polymer** A long-chain molecule made up of repeating units of monomers.

**precipitate** An insoluble solid formed from a reaction taking place in aqueous solution.

**product** A substance made in a chemical reaction.

**proton** A positively charged sub-atomic particle found in the nucleus of atoms.

**reactant** A substance at the start of a chemical reaction.

**reaction rate** A measure of change in concentration of a reactant with time. The greater the change the faster the reaction.

**redox reaction** A reaction in which both reduction and oxidation are taking place.

**reducing agent** A substance that is capable of reducing another substance.

**reduction** The addition of electrons to a substance, or the removal of oxygen.

**relative atomic mass ($A_r$)** The weighted mean mass of an atom of an element relative to one-twelfth $\left(\frac{1}{12}\right)$ the mass of an atom of carbon-12.

**relative formula mass ($M_r$)** The sum of the relative atomic masses of all the atoms present in the formula of a substance.

**reversible reaction** A reaction that can take place in both directions. It is represented by the use of the symbol $\rightleftharpoons$ in the equation.

**rusting** A chemical reaction between iron and oxygen and water from the air in which rust (hydrated iron(III) oxide) is formed.

**salt** A substance formed when the hydrogen ions in an acid are replaced by either metal ions or ammonium ions.

**saturated hydrocarbon** A hydrocarbon containing only single bonds.

**slag** The waste material produced in the blast furnace during the production of iron.

**theoretical yield** The maximum amount of a product that could be formed from a given amount of a reactant.

**thermal decomposition** Breaking down by using heat.

**titration** A method to determine the exact volume of one solution that will react with a given volume of another solution.

**unsaturated hydrocarbon** A hydrocarbon containing a carbon–carbon double bond.

**yield** The amount of product obtained in a chemical reaction. It is usually expressed as a percentage of the maximum possible amount of product that can be made from a given amount of reactant.

# Index